Arduino 101

A Technical Reference to Setup and Program Arduino Zero, Nano, Due, Mega and Uno Projects

OBAKOMA G. MARTINS

Copyright

Copyright©2020 Obakoma G. Martins

All rights reserved. No part of this book may be reproduced or used in any manner without the prior written permission of the copyright owner, except for the use of brief quotations in a book review.

While the advice and information in this book are believed to be true and accurate at the date of publication, neither the authors nor the editors nor the publisher can accept any legal responsibility for any errors or omissions that may be made. The publisher makes no warranty, express or implied, with respect to the material contained herein.

Printed on acid-free paper.

Table of Contents

Copyright i

CHAPTER ONE 1

1.0. Introduction 1

1.1 Installing the IDE 2

1.2. Setting up the Arduino board 7

1.3 Uploading and Running a Blink Sketch 16

1.4 Saving a Sketch 19

1.5 Attach the Parts 20

1.6 Upload the Blink Sketch 21

1.7. Using a 32-bit Arduino 23

CHAPTER TWO 25

2.1 Arduino Sketch Structure 25

The set up Function 25

The Loop Function 25

2.2 Variables 26

Declaring Variables 26

Using Variables 27

2.3. Using floating-point numbers 28

2.4. Using a group of values or Arrays 31

Creating an Array 31

Getting to an Array 31

2.5. Using an Arduino String 33

String Character Arrays 33

2.6. Converting a String to a Number on Arduino 36

2.7. Taking actions based on conditions 38

2.8. Returning More than one value from a function 39

Returning multiple values using an array 41

CHAPTER THREE 43

3.1. Mathematical Operations 43

3.2. Increment and Decrement of values 44

3.3. Finding the Remainder after dividing two values 45

3.4. Determining the absolute value 46

3.5. Constraining a Number to a range of values 47

3.6. Obtaining the maximum and minimum value 47

3.7. Raising a number to a power 50

3.8. Taking the Square root 50

3.9. Rounding Floating point number up and down 51

3.10. Using Trigonometric functions 52

3.11. Generating Random Numbers 52

3.12. Reading and Setting a Bit 54

3.13. Shifting a Bit 54

3.14. Extracting a high and low bytes in an integer or long 55

CHAPTER FOUR 56

4.0. Serial Communication 56

4.1 Sending information from Arduino the computer 56

4.2 Sending Formatted Text and Numeric Data from Arduino ... 58

4.3. Receiving Serial Data in Arduino 59

4.4 Receiving Multiple Text Fields in a Single Message in Arduino
.... 61

4.5. Sending Binary Data from Arduino 63

4.7. Receiving Binary Data from Arduino on a Computer 64

4.8. Sending Binary Values from Processing to Arduino 66

4.9. Sending the Value of Multiple Arduino Pins 68

4.10. Logging Arduino Data to a File on Your Computer 70

4.11. Sending Data to Two Serial Devices at the Same Time 74

4.12 Using Arduino with Raspberry Pi 4 76

CHAPTER FIVE 79

5.1 Using a Switch with the Arduino 79

5.2 Using a Switch without external resistors 82

5.3. Reliably Detecting when a switch is pressed (Debounce method) 84

5.4. Determining how long a switch is pressed 86

5.5. Reading a Keypad 88

5.6. Reading analog values 92

5.7. Changing the Range of Values 95

5.9. Measuring voltages greater than 5V 97

CHAPTER SIX 101

6.1 Detecting Movement 101

6.2. Detecting Light 104

6.3. Measuring Distance Precisely 106

6.4. Detecting vibration 110

6.5. Detecting Sound 112

6.6. Measuring Temperature 115

6.7. Reading RFID (NFC) tags 117

6.8. Using Mouse Button Control 122

6.9. Getting location from a GPS 125

6.10. Reading Acceleration 129

CHAPTER SEVEN 134

7.0. Visual Output 134

7.1. Connecting and Using LED 134

7.2. Adjusting the brightness of an LED 137

7.3. Adjusting the color of an LED 138

7.4. Sequencing Multiple LEDs 140

7.5. Controlling an LED matrix through multiplexing 144

7.6. Using an analog panel meter as display 146

CHAPTER EIGHT 149

8.1. Controlling Rotational position with a servo 149

8.2. Controlling Servo Rotation with a Potentiometer 151

8.3. Controlling a Brushless motor 153

8.4. Controlling a solenoid 155

8.5. Driving a brushed motor using a transistor 157

8.6. Controlling a Unipolar stepper 160

8.7. Controlling a Bipolar stepper 163

CHAPTER NINE 167

9.1. Playing Tones 167

9.2. Generating More Than One Simultaneous Tone 168

9.3. Controlling MIDI 171

CHAPTER TEN 175

10.1 Responding to an infrared remote control 175

10.2. Controlling a Digital Camera 178

10.3. Using millis to determine duration 180

10.4. Creating Pauses in your sketch 181

10.5. Arduino as a clock 182

10.5. Using a real-time clock 187

CHAPTER ELEVEN 190

11.1. Connecting to an Ethernet network 190

11.2. Using Arduino as a webserver 191

11.3. Sending Twitter messages 194

11.5. Publishing Data to an MQTT broker 196

11.6. Using built-in Libraries 198

11.7. Installing a third-party library 199

CHAPTER TWELVE 201

12.1. Arduino Build Process 201

12.3. Store and Retrieve Values 203

12.4. Storing Data permanently on EEPROM memory 205

12.5. Periodic Interrupt 206

12.6. Changing a Timer's PWM frequency 208

12.7. Counting Pulses 211

12.8. Reducing Battery Drain 212

12.9. Uploading Sketches using a programmer 213

12.10. Replacing Arduino Bootloader 215

About the Author 217

CHAPTER ONE
1.0. Introduction

Arduino is an open-source stage utilized for developing electronic projects. Arduino comprises both a physical programmable circuit board (regularly alluded to as a microcontroller) and a bit of programming, or IDE (Integrated Development Environment) that functions on your PC, used to compose and transfer PC code to the physical board. The Arduino stage has gotten very mainstream with individuals simply starting out with gadgets, and for valid justifications. Dissimilar to most past programmable circuit boards, the Arduino needn't bother with a different piece of equipment (called a programmer) in order to stack new code onto the board, you can essentially utilize a USB cable. Furthermore, the Arduino IDE utilizes an improved rendition of C++, making it simpler to figure out how to program. At long last,

Arduino gives a standard structure factor that breaks out the elements of the micro-controller into a more open bundle. Arduino is a programmable circuit board that can be coordinated into a wide assortment of makerspace ventures both straightforward and complex. Due it's adaptability and minimal effort, Arduino has become an exceptionally mainstream decision for producers and beginners hoping to make intelligent hardware projects. With the Arduino board, you will be able to read several inputs such as light on a sensor, a finger on a button, or a tweet and transform it into an output, which might be activating a motor, turning on a LED, distributing something on the web.

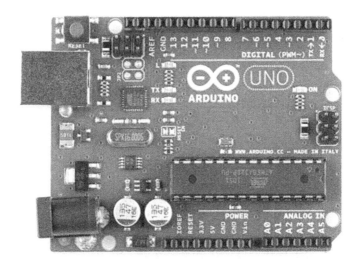

An Arduino Board

1.1 Installing the IDE

Perhaps you've just purchased an Arduino, and you're anxious to explore it, however you might be wondering where to start. If you must carry out anything useful on the Arduino, you have to input certain codes into it by installing the Arduino IDE on another computer. The IDE converts the codes you input into instructions that the Arduino can comprehend. When you compose the codes into the IDE, they are sent to the Arduino through a USB cable. The Arduino IDE software download and installation is free and easy to do.

- Visit http://www.arduino.cc/en/main/software to download the most recent Arduino IDE version for your computer's OS. There are forms meant for Windows, Mac, and Linux frameworks. When you're at the download page, click on the "Windows Installer" option to provide you with the quickest installation method.

-Store the .exe document to your hard drive and launch it.

- Press the button to confirm that you agree with the licensing statements.

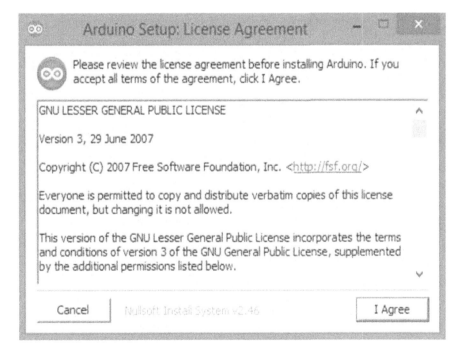

- Select your most preferred options to install and press Next

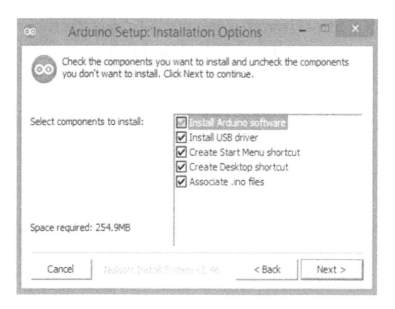

- Pick the folders you want the program to be installed to, then press install.

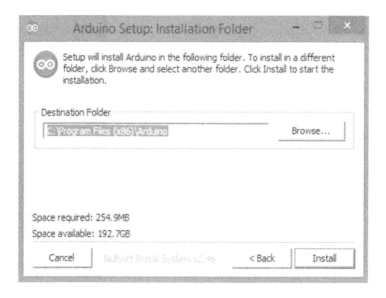

- Hold on for the program to complete the installation and close it.

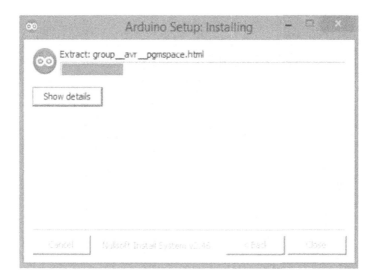

- Now locate the Arduino shortcut on your computer and open it. The IDE will be opened and the code editor will be displayed.

- Configuring the Arduino IDE.

The next task is to ensure the software is set up for your specific Arduino board. Click on "Tools" drop-down menu, and locate "Board". Another menu will show up, you can choose from a rundown of Arduino models from it.

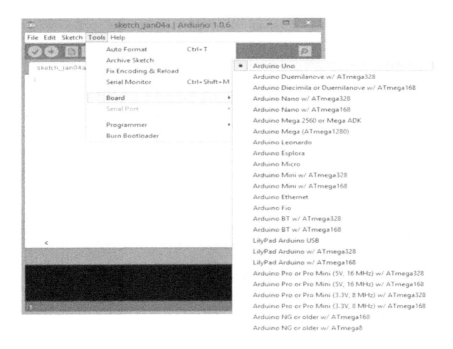

- Exploring the Arduino IDE

If the need to do this arises, take a moment to peruse the various menus in the IDE. There is a decent variety of model programs that accompany the IDE in the "Examples" menu. These will assist you with becoming familiar with your Arduino immediately without doing much research.

1.2. Setting up the Arduino board

What is on the Arduino board?

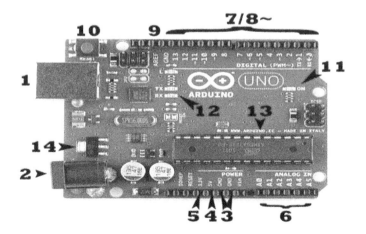

There are numerous varieties of Arduino boards that can be utilized for various purposes. A number of the boards appear to be different, however most Arduinos have most of these parts in common:

- **Power (USB/Barrel Jack)**

Each Arduino board has to be connected to a power outlet. The Arduino UNO can be supplied power from a USB link originating from your PC or a wall socket that ends in a barrel jack. In the image above, the USB port is labelled as (1) and the barrel jack is labelled as (2). In addition, it is through the USB connection that you will input the codes into your Arduino board. However, don't connect to a power supply that is larger than 20 Volts, because this will overpower and damage your Arduino. The suggested voltage for most Arduino models is somewhere in the range of 6 and 12 Volts

- **Pins (5V, 3.3V, GND, Analog, Digital, PWM, AREF)**

The pins on your Arduino are where you associate wires to build a circuit (most likely in conjunction with a breadboard and some wire. Usually, they have dark plastic 'headers' that enables you to simply plug a wire directly into the board. The Arduino has various types of pins, each of which is marked on the board and utilized for numerous functions.

- **GND (3):** abbreviated form for 'Ground'. There are numerous GND pins on the Arduino, any of which can be utilized to ground your circuit.
- **5V (4) and 3.3V (5):** the 5V pin produces 5voltels of power, while the 3.3V pin supplies a power of 3.3volts. The vast majority of the basic components utilized with the Arduino run on 5 or 3.3 volts.
- **Analog (6):** The zone of pins under the 'Analog In' name (A0 through A5 on the UNO) are Analog In pins. These pins can analyze the signal from an analog sensor (similar to a temperature sensor) and convert it into a computerized value that is human readable.
- **Digital (7):** on the other side from the analog pins are the digital pins (0 through 13 on the UNO). These pins can be utilized for both computerized input (such as confirming whether a button was pushed) and other digital outputs such as powering a LED.
- **PWM (8):** You may have seen the tilde (~) close to a portion of the advanced pins (3, 5, 6, 9, 10, and 11 on the UNO). These pins function as digital pins, however can likewise be utilized for something many refer to as Pulse-Width Modulation (PWM). These pins can recreate analog outputs.
- **AREF (9):** represents Analog Reference. Mostly, you might not use this pin. It is employed to set an outer reference voltage (between 0 and 5 Volts) as the upper limits for analog input pins.

- **Reset Button**

Similar to any electronic device you know, the Arduino has a reset button (10). Pushing it will quickly interface the reset pin with the ground and restart any code that is present on the Arduino. This can be extremely valuable if your code doesn't repeat, and you need to test it multiple times. However, resetting it doesn't fix any problem.

- **Power LED Indicator**

Just underneath, and to the right side of "UNO" on your circuit board, there's a small LED close to the word 'ON' (11). This LED should illuminate anytime you connect your Arduino into a power outlet. In the event that this light doesn't turn on, there's a decent possibility something isn't right. You should recheck your circuit.

- **TX RX LEDs**

TX is an abbreviation for transmit, RX represents Receive. These markings show up a lot in electronics to demonstrate the pins answerable for sequential communication. There are two spots on the Arduino UNO where TX and RX show up (once by advanced pins 0 and 1, and a second time close to the TX and RX pointer LEDs (12)). These LEDs will provide us with some pleasant visual signs at whatever point our Arduino is getting or sending information.

- **Main IC**

The black component with all the metal legs is an Integrated Circuit (13). Consider it the brain of our Arduino. This IC on the Arduino is somewhat different depending on the board type, yet is typically from the ATmega line of IC's from the ATMEL organization. This can be important because you need to know the IC type (alongside your board type) before entering a new program from the Arduino programming. Details about this are on the top side of the integrated circuit. If you intend to know more about the differences between the IC's, the datasheet may help you out.

- **Voltage Regulator**

The voltage regulator (14) isn't really something that is friendly on the Arduino. However, it is possibly helpful to know that it is there and what it's for. Just as the name implies, it regulates the quantity of voltage that is allowed into the Arduino board. It will dismiss an additional voltage that may damage the circuit. In addition, the regulator itself has limits, so don't power your Arduino to anything greater than 20 volts. Now that you're aware of the various components of an Arduino board, let's discuss the set up.

Step 1: Download and Install the IDE

The IDE is available for download on the official Arduino site. Since the Arduino employs a USB to serial converter, which enables it to interact with the host PC, the Arduino board is viable with most PCs that have a USB port. That being said, you will require the IDE first. Fortunately, the Arduino architects have produced numerous versions of the IDE for various operating systems, including Windows, Mac, and Linux. Therefore, you should download the IDE that is appropriate for the Operating System your computer has. After downloading it, install and enable all the options.

Step 2: Get the Arduino COM Port Number

Next, you'll have to interface the Arduino Uno board to the PC. You can do this by means of a USB connection. The USB supplies the board with 5V up to 2A. When you connect the Arduino, the operating system ought to identify the board as a conventional COM port. When the computer recognizes the board, you should look for the port number assigned to it. The simplest method to do this is to type "device manager" into Windows Search and press Device Manager when it shows.

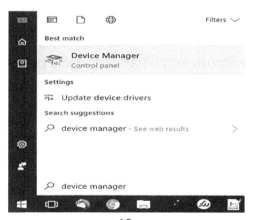

In the Device Manager window, search for a device under "Ports (COM and LPT)", and the Arduino might probably be the main gadget on the list. In my Device Manager, the Arduino appears as COM7. However, the computer won't always identify the Arduino automatically. In case the PC doesn't recognize the Arduino, uninstall the driver, remove the Arduino, insert the Arduino again, locate the unrecognized device, right click "Update driver", and afterward click "Search consequently". This should fix the issue almost all the time.

Windows can be a genuine torment at times with COM ports, as it can mystically change their numbers between connections. Sometimes your Arduino might be on port 7, however on another day, Windows may transfer it to an alternate port number. This mostly happens when you interface other COM ports to your computer. Thus, if you can't locate your Arduino on the port that you normally use, simply go to your Device Manager and check what port it is, and update your driver if need be.

Step 3: Configure the IDE

Now that we have fixed the port that the Arduino is on, we can now go ahead to load the Arduino IDE, then program it to use a particular device and port. Firstly, load the IDE. When it's loaded, go to Tools > Board > Arduino Uno. Notwithstanding, if you are using a board that's different that is different from the Arduino, you have to select the correct board.

Inform the IDE about the board you're using.

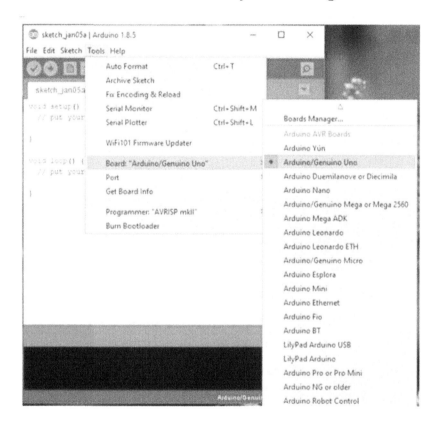

At this point, you should inform the IDE about the COM port the Arduino is on. To do this, go to Tools > Port > COM7. Clearly, if your Arduino is on another port, select that port.

- **Preparing an Arduino Sketch with the IDE**

When you have effectively installed Arduino IDE, you can start coding. The source code documents for Arduino are called sketches. A sketch stands for a name that Arduino uses for a program. It's the unit of code that is uploaded to and operated on an Arduino board. The Arduino programming language is based on C/C++ and it is almost the same. Open the Arduino IDE and a blank sketch will show up on your screen

The sketch is separated into two program parts: set up and loop. In set up, you make fundamental hardware and software configurations. This aspect of the code runs only once. For instance, if we are driving a LED, we can program the computerized I/O pin we have our LED attached to as an output pin.

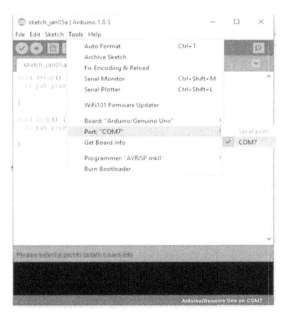

Since we have finished setting up the Arduino, we can compose the main body of the code. This will fall under the loop aspect and will rehash repeatedly except it is instructed otherwise or if power is taken out from the Arduino.

We will then input these commands if we want to flash the LED on and off

1) Turn LED On

2) Wait ½ of a second (500 milliseconds)

3) Turn LED Off

4) Wait ½ of a second

5) Repeat

Since the code we compose is inside the loop function, the Arduino will consequently repeat the code.

1.3 Uploading and Running a Blink Sketch

The kind of sketch we intend to use here is called Blink. Click in the Arduino window. From the menu bar, press File→Examples→01.Basics→Blink.

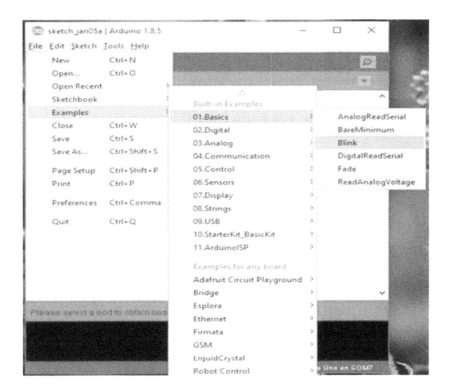

Connect the Arduino board to your PC via the USB. After interfacing the Arduino, follow these steps to upload the sketch:

1) Select the type of board you're using, that is the target board

The board you select informs the Arduino IDE which Arduino board you are uploading to. If you have another board other than the Arduino Uno, select that board

2) Choose the serial port the board is associated with

Now that we have informed the Arduino software about the type of board you are interacting with and which serial port connection it is using, you can upload the Blink sketch.

Press the Verify button. This checks whether the inputted codes fit well. This doesn't actually imply that your code will do what you are expecting, however it verifies that the language structure is written in a manner Arduino can comprehend. You should notice a progress bar and the content Compiling Sketch for a couple of seconds followed by the text Done compiling after the concluding the process

Then, if the sketch compiled successfully, you can press the Upload button near the verify button. A progress bar shows up, and you see numerous activities on your board from the two LEDs marked RX and TX. These imply that the Arduino is sending and receiving information. After a while, the RX and TX LEDs quit blinking, and a Done Uploading message shows up at the base of the window.

You should see the LED stamped L blinking intermittently: on for a second, off for a second

1.4 Saving a Sketch

The sketches included with the Arduino IDE are 'read-only'. This means that you can upload them to an Arduino board, however if you change them, you can't save them as a similar document. Thus, the main thing you have to do is save your own duplicate that you can change anyway you like.

From the File menu on the Arduino IDE, press 'Save As..' and afterward, save the sketch with the name 'MyBlink'.

A copy of the sketch has now been saved in your sketchbook. This implies anytime you want to use it, you can simply open it via the File → Sketchbook menu option.

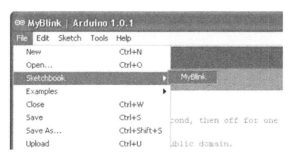

An Easy Arduino Sketch: Blink an LED

Required components;

- Arduino Uno Board
- Breadboard – half size
- Jumper Wires
- USB Cable
- LED (5mm)
- 220 Ohm Resistor

1.5. Attach the Parts

You can create your Arduino circuit by adhering to the manner in the breadboard image below or via the description. In the description, we will utilize a letter/number combination that alludes to the area of the component. If we focus on H19 for instance, that represents column H, row 19 on the breadboard.

Step 1 – Insert the black jumper wire into the GND (Ground) pin present on **the Arduino and afterward in the GND rail of the breadboard row 15**

Step 2 – This time, insert the red jumper wire into pin 13 on the Arduino and afterward the opposite end into F7 on the breadboard

Step 3 – Insert the LONG leg of the LED into H7

Step 4 – Insert the SHORT leg of the LED into H4

Step 5 – Twist the two legs of a 220 Ohm resistor and insert one leg in the GND rail around row 4 and other leg in I4.

Step 6 – Connect the Arduino Uno to your PC through USB link

1.6. Upload the Blink Sketch

This is a perfect time to upload the sketch to the Arduino and instruct it. The IDE has implicit example sketches that you can utilize to make it simple.

To gain entry into the blink sketch, you should go to File > Examples > Basics > Blink

You should have a fully coded blink just like the image below at this point.

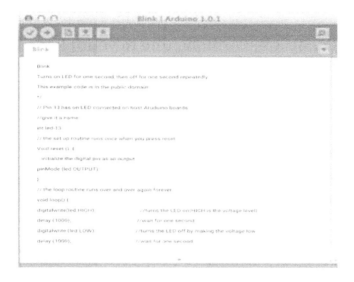

Next, you have to press the verify button (check mark) that is situated in the upper left of the IDE box. This will compile the sketch and search for errors. When it shows "Done Compiling", it can be uploaded. Press the upload button, represented by the forward arrow, to send the program to the Arduino board.

The implicit LEDs on the Arduino board will flash quickly for a couple of seconds and afterward the program will execute. If it goes as expected, the LED on the breadboard should flash intermittently.

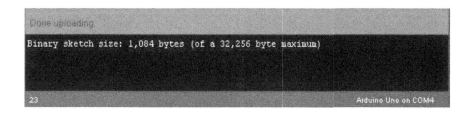

With that, you have just successfully completed your first Arduino project.

1.7. Using a 32-bit Arduino

The Arduino Due is a microcontroller board based on the Atmel SAM3X8E ARM Cortex-M3 CPU. It is the first Arduino board to be built based on a 32-piece ARM center microcontroller. It has 54 advanced input/output pins (of which 12 can be utilized as PWM yields), 12 simple information sources, 4 UARTs (hardware ports), a 84 MHz clock, a USB OTG cable, 2 DAC (digital to analog), 2 TWI, a power port, a SPI header, a JTAG header, a reset button and a delete button.

The board entails all that is expected to help the microcontroller; essentially interface it to a PC with a micro USB cable or force it with an AC-to-DC connector or battery to begin. The Arduino Due works well with virtually all Arduino shields that operate at 3.3V and are compatible with the 1.0 Arduino pinout. The Due follows the 1.0 pinout:

- **TWI:** SDA and SCL pins that are close to the AREF pin.

- **IOREF:** permits a connected shield with the best possible design to adjust to the voltage supplied by the board. This empowers shield compatibility with a 3.3V board like the Due and AVR-based sheets which work at 5V.

- A detached pin, saved for future use.

CHAPTER TWO

2.1 Arduino Sketch Structure

A typical Arduino sketch comprises two functions called set up and loop.

Open the Arduino IDE and select File → Examples → 01.Basics → BareMinimum to check these two functions. These functions can be accessed in a default new Arduino IDE window, so it isn't important to open the BareMinimum example sketch in another model of the IDE.

The set up Function

Commands in the setup function are run just once; each time you run the sketch. The program then begins to execute commands in a loop function. The sketch will start to operate after entering it into the Arduino. Launching the serial monitor window will reset the Arduino and cause it to run the sketch once more.

The Loop Function

Orders tuned in loop operation will run reliably from start to finish and a while later back to the top. In case the circle work contains two orders, the primary order will be executed, afterwards the subsequent order, the main order once more, and like that in a circle.

The presence of the loop is basic in the sketch, whether or not it is unfilled, in light of the fact that without it the microcontroller on the Arduino board will endeavor to execute whatever it finds next in memory after the orders in the set up work have been executed. The microcontroller will attempt to complete whatever it finds in memory as a guidance; this is forestalled by the circle work by holding the program execution on top of it.

Parts of a Sketch

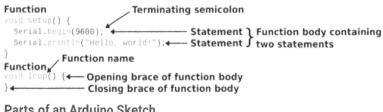

Parts of an Arduino Sketch

Statements are lines of code that are executed when you operate the program. Every statement is halted with a semicolon.

2.2 Variables

A variable is a method of naming and putting a value for future use by the program, for example, data obtained from a sensor or a moderate value utilized in a computation.

Declaring Variables

You have to declare a variable before you use it. Declaring a variable implies that you are characterizing its type and alternatively, setting an initial value(this is called initializing the variable).

int inputVariable1;

int inputVariable2 = 0; //both are correct

You ought to consider the size of the numbers they intend to store in picking variable types. Variables will turn over when the attached value surpasses the space allotted to store it.

Using Variables

After declaring a variable, you can characterize them by setting the variable equivalent to the value one intends to store with the assignment operator (single equal sign). The program operator instructs the program to enter whatever is on the right side of the equal sign into the variable on the left side.

inputVariable1 = 7;   // sets the variable named inputVariable1 to 7

inputVariable2 = analogRead(2); // sets the variable named inputVariable2 to the

          // (digitized) input voltage read from analog pin #2

Examples

  int lightSensVal = 1234;

  char currentLetter = 'a';

  unsigned long speedOfLight = 186000UL;

  char errorMessage[] = "choose another option"; // see string

After setting a variable by assigning it a value, you can ascertain its value to check whether it satisfies certain requirements, or you can utilize its value directly. For example, these tests whether the inputVariable2 is less than 100, afterwards set a delay based on inputVariable2 which has 100 as its minimum

```
if (inputVariable2 < 100)
{
  inputVariable2 = 100;
}
delay(inputVariable2);};
```

2.3. Using floating-point numbers

The datatype for floating-point numbers, that is, numbers that have a decimal point. Floating-point numbers are regularly used to approximate analog and persistent values since they have a higher resolution than whole numbers. Floating-point numbers can range from large to low such as from 3.4028235E+38 to - 3.4028235E+38. Floating-point numbers are seen in their memory location as 2 bits (4 bytes) of data.

Syntax

float var = val;

Parameters

var: variable name.

val: the value you assign to that variable.

This example depicts all the three useful operations with variables. It inspects the variable (if (inputVariable2 < 100)), it puts the variable in place on the off chance that it passes the test (inputVariable2 = 100), and it employs the value of the variable as an input parameter for the delay function (delay(inputVariable2))

Example Code

float myfloat;

float sensorCalbrate = 1.117;

int x;

int y;

float z;

x = 1;

y = x / 2;   // y now contains 0, ints can't hold fractions

z = (float)x / 2.0; // z now contains .5 (you have to use 2.0, not 2)

It is important you know that when doing math with floats, you have to include a decimal point, else it will be treated as an integer.

The float data type entails about 6-7 decimal digits of accuracy. This implies the entire number of digits, not the number to the right side of the decimal point. In contrast to other platforms, where you can be provided with more precision by utilizing a double(up to 15 digits), on the Arduino, a float has a similar size as that of a double.

Floating-point numbers are not precise, and may produce random results when analysed. For instance 6.0/3.0 may not equal 2.0. Instead, you ought to check the absolute value of the distinction between the numbers is less than some small number.

Changing from floating point to integer may provide you with truncated results;

```
float x = 2.9; // A float type variable
int y = x; // 2
```

If, instead, you want to round off during the conversion process, you need to add 0.5:

float x = 2.9;

int y = x + 0.5; // 3

or use the round() function:

float x = 2.9;

int y = round(x); // 3

Also, floating-point math is less quicker than integer math in carrying out calculations, so ought to be avoided if, for instance, a loop needs to operate at maximum speed for a basic function.

2.4. Using a group of values or Arrays

An array means a collection of variables that have index numbers. The manner in which arrays in the C++ programming language Arduino portrays are written look complicated, but simple arrays are relatively easier to use.

Creating an Array

You can create (declare) an array through any of these methods:

```
int myInts[6];
int myPins[] = {2, 4, 8, 3, 6};
int mySensVals[6] = {2, 4, -8, 3, 2};
char message[6] = "hello"
```

You can create an array without introducing it as in myInts.

In myPins, an array is declared without picking a size. The compiler gives an estimate of the entities and makes an array of the appropriate size. Moreover, you can in state and measure your exhibit, as in mySensVals.

Getting to an Array

Clusters can be zero field, that is, taking an example from the exhibit instatement over, the principal component of the exhibit is at record, thus mySensVals[1] == 4, etc.

It likewise implies that in an array with ten components, index nine is the last component. Therefore;

```
int myArray[10]={9, 3, 2, 4, 3, 2, 7, 8, 9, 11};

// myArray[9]   contains 11

// myArray[10]  is invalid and contains irregular data (other memory address)
```

Therefore, you should be careful when accessing arrays. Accessing beyond the terminal of an array (utilizing a file number greater than the created array size - 1) is perusing from memory that is being used for different purposes. Reading from these areas won't deliver much other than yield invalid data. Writing in a non-sequential manner to different memory locations is not such a good idea and might give unpleasant results such as crashes or program malfunction. This can also be a difficult issue to troubleshoot.

In contrast to BASIC or JAVA, the C++ compiler does not verify whether the array access is within legitimate limits of the array size that you have created.

To assign a value to an array:

```
mySensVals[0] = 10;
```

To retrieve a value from an array:

```
x = mySensVals[4];

};
}
```

2.5. Using an Arduino String

Texts are stored with strings. They can be utilized to show text on a LCD or in the Arduino IDE Serial Monitor window. Strings are likewise helpful when it comes to storing the users' input.

There are two types of strings in Arduino programming;

- Array of characters, which are equivalent to the strings utilized in C programming.

- The Arduino String, which enables us to use a string object in a sketch.

String Character Arrays

Specifically, what we'll look into is a string that is an array of characters of the type char. An array is a sequential series of a similar sort of variable stored in memory. Likewise, a string is an exceptional cluster that has one extra component toward the terminal of the string, which reliably has an estimation of 0 (zero). This is known as "null termination of a string"

String Character Array Example

The following examples will guide you on how to create a string and print it to the monitor windows.

Example

```
void setup() {
    char my_str[6]; // an array big enough for a 5 character string
    Serial.begin(9600);
    my_str[0] = 'H'; // the string consists of 5 characters
    my_str[1] = 'e';
    my_str[2] = 'l';
    my_str[3] = 'l';
    my_str[4] = 'o';
    my_str[5] = 0; // 6th array element is a null terminator
    Serial.println(my_str);
}
void loop() {

}
```

Also, these examples will inform us about what a string consists of; a character array with printable characters and 0 as the last element of the array to indicate this is the point the string finalizes. The string will be outputted on the Arduino IDE Serial Monitor window through the Serial.println() and passing the name of the string.

Likewise, there are ways to write such strings in a more convenient way;

```
void setup() {
    char my_str[] = "Hello";
    Serial.begin(9600);
    Serial.println(my_str);
}
void loop() {
```

In this sketch, the compiler estimates the size of the string array and consequently null terminates the string with a zero. An array that possesses six letters in length and comprises of five characters followed by a zero is created in the same way as the previous sketch.

- Manipulating String Arrays

We can adjust a string array in a sketch, for instance;

```
void setup() {
    char like[] = "I like coffee and cake"; // create a string
    Serial.begin(9600);
    // (1) print the string
    Serial.println(like);
    // (2) delete part of the string
    like[13] = 0;
    Serial.println(like);
    // (3) substitute a word into the string
    like[13] = ' '; // replace the null terminator with a space
    like[18] = 't'; // insert the new word
    like[19] = 'e';

    like[20] = 'a';
    like[21] = 0; // terminate the string
    Serial.println(like);
}
void loop() {
}
```

Result

I like coffee and cake

I like coffee

I like coffee and tea

The adjustment is made to the sketch in this way:

- You first create and print the string.

- Then you shorten the string probably by replacing a character in the string with the null terminating zero.

- Changing words in the string. The words to be changed are first replaced with a null terminating zero, then you enter the new word.

2.6. Converting a String to a Number on Arduino

In some cases, a number is entered as a string. To use it for any numerical operation, we need to change the string to number, this can be done manually. Here's how to go about it, firstly, you have to know ASCII (American Standard Code for Information Interchange) characters and their decimal value.

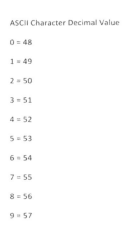

ASCII Character Decimal Value

0 = 48
1 = 49
2 = 50
3 = 51
4 = 52
5 = 53
6 = 54
7 = 55
8 = 56
9 = 57

Numbers are saved in character format inside the string. So to get the decimal value of each string component, we need to subtract it with the decimal value of character '0.' Let's make this reasonable through an example;

Example: Program to Manually Convert a String to an Integer

```
void setup()
{
  Serial.begin(9600);
}
void loop()
{
  String num = "1234";
  int i, len;
   result=0;
  Serial.print("Number: ");
  Serial.println(num);
  len = num.length();
  for(i=0; i<len; i++)
  {
    result = result * 10 + ( num[i] - '0' );
  }
  Serial.println(result);
}
```

The program code is written inside the brace brackets of the main function. Within the loop, we initially identify and declare the various variables alongside their data types. Variables i, len and result are regarded as of integer data type. The outcome variable is instated to zero. The serial.print() work is then summoned to show the message "number" on the terminal screen. Here, the string is an array of characters highlighted by num. At that point, we estimate the length of the string via the length() function. Afterwards, we loop through the string and convert the string into decimal worth. Finally, the string will be converted to a number and displayed on the screen.

2.7. Taking actions based on conditions

The if() statement is the most fundamental of all programming control structures. It enables you to get something going or not, contingent upon whether a given condition is valid or not. It would appear this way:

if (someCondition) {

  // do stuff if the condition is true

}

There is a typical variation referred to as if-else, and it looks like this

if (someCondition) {

  // do stuff if the condition is true

```
} else {
    // do stuff if the condition is false
}
```

Likewise, there's the else-if, where you can check a subsequent condition if the first statement is false:

```
if (someCondition) {
    // do stuff if the condition is true
} else if (anotherCondition) {
    // do stuff only if the first condition is false
    // and the second condition is true
}
/and the subsequent condition is valid
}
```

2.8. Returning More than one value from a function

Programmers are typically in the pursuit of approaches to return multiple values from a function. However, C and C++ don't permit this function directly. Nevertheless, this can be done with a bit of intelligent programing

There are several ways to get this done, some of which are:

- By using pointers.

- By using Arrays.

For instance, consider an operation where the activity is to locate the greater and smaller number of the two. We could compose multiple functions. The primary issue is the difficulty of calling more than one function since we have to restore multiple functions, having more number of lines of code to be inputted.

Returning multiple values using pointers:

Transmit the argument with their address and make changes in their values utilizing pointer. So that the values are changed into the original argument.

```c
// Modified program using pointers
#include <stdio.h>
// add is the short name for address
void compare(int a, int b, int* add_great, int* add_small)
{
    if (a > b) {
        // a is stored in the address pointed
        // by the pointer variable *add_great

        *add_great = a;
        *add_small = b;
    }

    else {
        *add_great = b;
        *add_small = a;

    }
}
// Driver code
```

```
int main()
{
    int great, small, x, y;
    printf("Enter two numbers: \n");
    scanf("%d%d", &x, &y);
    compare(x, y, &great, &small);
    printf("\nThe greater number is %d and the"
        "smaller number is %d",
        great, small);
    return 0;
}
```

Output:

Enter two numbers:

5 8

The greater number is 8 and the smaller number is 5

Returning multiple values using an array

This is effective only when the items you're returning are similar. When an array you identify an element as an argument, then its base address is passed to the function, therefore any changes you make to the duplicate array, it is included in the main array.

Below is the program to return multiple values using array i.e. store greater value at arr[0] and smaller at arr[1].

```c
// Modified program using array
#include <stdio.h>
// Store the greater element at 0th index
void findGreaterSmaller(int a, int b, int arr[])
{

    // Store the greater element at

    // 0th index of the array

    if (a > b) {

        arr[0] = a;
        arr[1] = b;

    }
    else {
        arr[0] = b;
        arr[1] = a;

    }
}
// Driver code
int main()
{

    int x, y;

    int arr[2];

    printf("Enter two numbers: \n");

    scanf("%d%d", &x, &y);

    findGreaterSmaller(x, y, arr);

    printf("\nThe greater number is %d and the"

        "smaller number is %d",

        arr[0], arr[1]);

    return 0;
}
```
Output:

Enter two numbers:

5 8

The greater number is 8 and the smaller

CHAPTER THREE

3.1. Mathematical Operations

The Arduino can also carry out arithmetic operations. To do this, we will employ certain arithmetic operators on Arduino

Addition

```
int a = 2;
int b = 3;
int sum;

sum = a + b;
```

Subtraction

```
int a = 5;
int b = 3;
int subtract;

subtract = a - b;
```

Multiplication

```
int multiply;

multiply = 6 * 3;
```

Division

Int divide;

Divide = a/ b;

3.2. Increment and Decrement of values

The increment and decrement functions are Arduino mathematical operators that increase or decrease an integer variable by a value of one. This is helpful in some types of loops.

Two potential methods of increment operation are:

Variable_Name++ and Variable_Name --: As the '++' and '--' signs come after the variable name, it is an operation that happens after the increment/ decrement. This implies the variable is first utilized in the statement and incremented/ decremented after executing the statement.

++Variable_Name and--Variable_Name: As the signs come before the variable name, it is an operation that happens before the increment/decrement. This implies the variable is increased/decreased before the statement is executed.

```
int count = 0;

void setup() {
  Serial.begin(9600);
}

void loop() {
  Serial.print("Count Value: ");
  Serial.println(count++);
  delay(1000);
}
```

Increment operation

```
int count = 100;

void setup() {
  Serial.begin(9600);
}

void loop() {
  Serial.print("Count Value: ");
  Serial.println(count--);
  delay(1000);
}
```

Decrement operation

3.3. Finding the Remainder after dividing two values

This is referred to as the Remainder/ Modulo Division, the remainder operator is used to find the remainder after dividing two numbers. The percentage sign (%) is used as the modulo operator.

For instance,

int result;

result = 11 % 3;

The result of this calculation will be the remainder of 11 divided by 3 which is 2.

3.4. Determining the absolute value

An element's absolute value can be determined by using an absolute operator. What an absolute function does is to provide you with the magnitude of its argument.

It gives you the positive values from any value that you inputted (either negative or positive). Mathematically it is:

$x = |x|$

```
if x =   300 then   |300| = 300,  or   abs(300)
== 300
if x =  -300 then   |-300| = 300,  or   abs(-300)
== 300
```

For instance;

int a = -5;

x = abs(a); x = abs(-5);

Result x is 5.

3.5. Constraining a Number to a range of values

The constraint function, just as its name implies, constrains a certain value or result to be within a particular range. Constrain(x, a, b).

Noting that;

x: the number you want to constrain, it could be any data type

a: the minimum value of the range, all data types

b: the maximum value of the range, all data types

With this we'll draw an outline that;

x: if x lies between a and b

a: if x is smaller than a

b: if x is higher than b

Example

sensVal = constrain(sensVal, 20, 200);

// limits range of sensor values to between 20 and 200

3.6. Obtaining the maximum and minimum value

To begin this function, we will bring the cpplinq library and afterward we declare the use of this library.

#include "cpplinq.hpp"

From that point onward, we will declare an array of whole numbers containing random

It is from these values that we will obtain the minimum and maximum values.

Serial.begin(115200);

int ints[] = {5,7,4,7,8,15,9,25,15,14,30,9,24,5,78,912,37,48, 980,200,201};

Before applying the operators, we have to transform the array to a range object. We then call the from_array function, and passing the array of integers as input

from_array(ints)

At that point, to get the maximum value of the array, we basically need to call the max operator. This administrator takes no contentions and will pick out the number with the maximum value from the array. i.e. max()

To get the minimum value from the array, we have to call the min function, which additionally takes no argument. This will display the integer with the minimum value from the array i.e. min()

In the two cases, we'll employ the >> operator between the call to the from_array function and the call to the cppling operator. In the two cases, the normal return esteem is a whole number.

```
int maxVal = from_array(ints)

        >> max();

int minVal = from_array(ints)

        >> min();
```

Lastly, we will print both values we have obtained from applying these operators.

```
Serial.println(maxVal);

Serial.println(minVal);
```

The final complete code;

```
#include "cpplinq.hpp"

using namespace cpplinq;

void setup() {

  Serial.begin(115200);

  int  ints[]  =  {5,7,4,7,8,15,9,25,15,14,30,9,24,5,78,912,37,48,980,200,201};

  int maxVal = from_array(ints)

        >> max();

  int minVal = from_array(ints)

        >> min();

  Serial.println(maxVal);
```

Serial.println(minVal);

}

}

3.7. Raising a number to a power

On the Arduino IDE, the function for determining the raised value of a number is the power function, Pow(). This function is used to raise a number to a certain exponent.

Function;

pow(base, exponent)

Base is the number (float), while exponent is the power to which the base is raised (float)

```
float i=0;
long result;

void(setup){
  Serial.begin(9600L);
}

void(loop){
  Serial.println("Base 2: ");
  for(i=0;i<50;i++){
    result=pow(2,i);
    Serial.println(result);
  }
}
```

3.8. Taking the Square root

The function to find the square root of an integer on the Arduino IDE is the square root function represented by sqrt(x), x could be any data type.

For instance

```
long number = 0;

void setup() {
  Serial.begin(9600L);
}

void loop() {
  number = sqrt(analogRead(IO_7));
  Serial.println(number);
  delay(1000);
}
```

3.9. Rounding Floating point number up and down

As stated, floating-point numbers are numbers that have a decimal point. Floating-point numbers normally used to convert to analog and continuous values or integers via the round function.

The round function is round (x) where x is a floating point number.

It is in this manner, float var = val;

Where var stands for variable name and val is the value you give to that variable.

However, if you intend to round off during the conversion process, you should add 0.5:

float x = 2.9;

int y = x + 0.5; // 3

or employ the round() function:

float x = 2.9;

int y = round(x); // 3

3.10. Using Trigonometric functions

Trigonometric functions are typically used for calculating the distance of moving objects or their angular speeds. Arduino can additionally carry out basic trigonometric functions such as sin, cos, tan, asin, acos, atan, they're operated when you input their prototypes.

```
double sin(double x); //returns sine of x radians
double cos(double y); //returns cosine of y radians
double tan(double x); //returns the tangent of x radians
double acos(double x); //returns A, the angle corresponding to cos (A) = x
double asin(double x); //returns A, the angle corresponding to sin (A) = x
double atan(double x); //returns A, the angle corresponding to tan (A) = x
```

Take for example:

double sine = sin(2); // approximately 0.90929737091

3.11. Generating Random Numbers

Random functions are generated using the random function usually within minimum and maximum range. The Data type is "Long"

- **Parameters**

Min: Stands for the lowest number of the random value, it is inclusive or proportional

Max: represents the highest number of the random value, exclusive

If it is relevant for a series of values generated by random() (the random function) to differ, use randomSeed() to begin the random number generator with an input that is modestly random, such as analogRead() on an unconnected pin.

Alternatively, it can be sometimes useful to utilize pseudo-random sequences that have exact repetition. This can be executed by calling randomSeed() with a fixed number, before initiating the random sequence. Here is an example of codes that generate random numbers and displays them

```
long randNumber;
void setup() {
  Serial.begin(9600);
  // if analog input pin 0 is unconnected, random analog
  // noise will cause the call to randomSeed() to generate
  // different seed numbers each time the sketch runs.
  // randomSeed() will then shuffle the random function.
  randomSeed(analogRead(0));
}
void loop() {
  // print a random number from 0 to 299
  randNumber = random(300);
  Serial.println(randNumber);
  // print a random number from 10 to 19
  randNumber = random(10, 20);
  Serial.println(randNumber);
  delay(50);
```

3.12. Reading and Setting a Bit

The value of a bit is 0 or 1, and you can read the value of a bit by using the BitRead function. The syntax is bitRead(x, n) where x is the number from which to read and n is which bit to read, with 0 as the least-significant bit.

Setting a bit requires using the bitSet function. This function by writing a 1 to a bit of a numeric variable. The sentence structure is bitSet(x, n)

Where x is the numeric variable whose bit to set. and n: which bit to set, starting at 0 for the least bit.

3.13. Shifting a Bit

In C++ programming, there are two bit shift operators,

- The left shift operator <<
- The right shift operator >>

The sentence structure here is:

variable << number_of_bits

variable >> number_of_bits

Parameter;

variable - (byte, int, long) number_of_bits

integer <= 32

Example:

```
int a = 5;      // binary: 0000000000000101

int b = a << 3;  // binary: 0000000000101000, or 40 in decimal

int c = b >> 3;  // binary: 0000000000000101, or back to 5 like we started with
```

Let's say you shift a value x by y bits (x << y), the leftmost y bits in x are lost, literally shifted out presence:

```
int a = 5;       // binary: 0000000000000101

int b = a << 14; // binary: 0100000000000000 - the first 1 in 101 was discarded
```

3.14. Extracting a high and low bytes in an integer or long

This code extracts bytes values from an integer value. For this task, we will take an integer value in hexadecimal format and proceed to extract all 4 bytes in different four variables. The way to implement this is to right shift value according to byte position and mask it for One byte value (0xff).

```c
/*C program to extract bytes from an integer (Hex) value.*/
#include <stdio.h>

typedef unsigned char BYTE;

int main()
{
    unsigned int value=0x11223344; //4 Bytes value

    BYTE a,b,c,d; //to store byte by byte value

    a=(value&0xFF);       //extract first byte
    b=((value>>8)&0xFF);  //extract second byte
    c=((value>>16)&0xFF); //extract third byte
    d=((value>>24)&0xFF); //extract fourth byte

    printf("a= %02X\n",a);
    printf("b= %02X\n",b);
    printf("c= %02X\n",c);
    printf("d= %02X\n",d);

    return 0;
}
```

Output
```
a= 44
b= 33
c= 22
d= 11
```

CHAPTER FOUR

4.0. Serial Communication

Serial Communication has a vital role in the communication system. All micro-controllers are built with a serial communication port. Serial communication grants a text-user interface (TUI). Serial ports in microcontrollers transmit data serially to laptop/PC and likewise receive data serially from laptop/PC. In this case, an Arduino board is used to interact with the laptop. Arduino boards have at least one serial port known as UART or USART to communicate. These ports interact with digital pin 0 and 1 of Arduino referred to as RX and TX pin as well as with the computer by using a USB.

4.1 Sending information from Arduino the computer

Sending information on the Arduino requires serial communication. Check the port your Arduino is appended to in your system, create a sketch to transmit the data through Arduino's serial communication and try setting up a simple Nodes.js application using the serialport package, which will receive your data. Not minding whether you are familiar with the Node or not, this is easy to carry out and can be done quickly.

- **Syntax**

Var SerialPort = require("serialport").SerialPort

Var serialPort = new SerialPort("/dev/ttyACM0", {

    Baudrate: 57600

});

Arduino is identified as ttyACM0 on Unix-based systems. With this line of code, you initiate a serial communication with your board.

Proceeding, you can register events such as data that come from the Arduino:

// it opens the connection and register an event 'data'

serialPort.on("open", function () {

  console.log('Communication is on!');

  // when your app receives data, this event is fired.

  // so you can capture the data and do what you need

  serialPort.on('data', function(data) {

    console.log('data received: ' + data);

  });

});

With that, your app has just received the data.

4.2 Sending Formatted Text and Numeric Data from Arduino

The data to the serial port can be printed in numerous formats. Following the sketch below, we obtain values in HEX, DEC and other forms.

Char chrValue = 65; // these are the starting value to print

Int intValue = 65;

Float floatValue = 65.0;

Void setup()

{

  Serial.begin(9600);

}

Void loop()

{

  Serial.println("charValue: ");

  Serial.println(chrValue);

  Serial.println(chrValue, DEC);

  Serial.println("intValue: ");

  Serial.println(intValue);

  Serial.println(intValue, DEC);

  Serial.println(intValue,HEX);

```
Serial.println(intValue, OCT);

Serial.println(intValue, BIN);

Serial.println("floatValue: ");

Serial.println(floatValue);

Delay(1000);  // delay a second between numbers

chrValue++;   //to the next value

intValue++;
}
```

4.3. Receiving Serial Data in Arduino

In this case, we intend to receive data on Arduino from a computer or another serial device. The Serial communication feature in Arduino is also useful here. It's relatively easier to transmit 8-bit values (chars and bytes) because the Serial function uses 8-bit values. The sketch receives a digit (single character 0 through 9) and blinks the LED on pin 13 at a rate equivalent to the digit value it receives.

```
*/

Const int ledpin = 13;

Int blinkRate = 0;

Void setup()
{
```

```
  Serial.begin(9600);
  pinMode(ledpin, OUTPUT);
}
Void loop()
{
  If ( Serial.available())
  {
    Char ch = Serial.read();
    If( ch >= '0' && ch <= '9')
    {
      blinkRate = (ch – '0');
      blinkRate = blinkRate * 100;
    }
  }
Blink();
}
Void blink()
{
```

digitalWrite(ledpin, HIGH);

delay(blinkRate);

digitalWrite(ledpin, LOW);

delay(blinkRate);

}

View rawReceiving serial data in Arduino.ino

Upload this sketch and use the serial monitor to send a message. Open the serial monitor and input a digit in the text box of the window. Press send and the inputted character will be sent, you should see the blink rate change.

4.4 Receiving Multiple Text Fields in a Single Message in Arduino

The sketch for this function accumulates values, but here, each value is inputted into an array. This array has to be large enough to hold all the fields. A character different from a digit or comma initiates the printing of all the values that have been stored in the array.

The syntax are shown below;

```
Else if (ch == ',') // comma is our separator, so move on to the next field
{
If(fieldIndex < NUMBER_OF_FIELDS-1)
fieldIndex++;
// increment field index
}
Else
{
Serial.print( fieldIndex +1);
Serial.println(" fields received:");
For(int i=0; I <= fieldIndex; i++)
{
Serial.println(values[i]);
Values[i] = 0; // set the values to zero, ready for the next message
}
fieldIndex = 0; // ready to start over
}
```

An alternative method is to use a library called TextFinder, which is accessible from the Arduino Playground. The basic function of a TextFinder is to extract information from a web stream, but it functions effectively with serial data. This sketch utilizes a TextFinder to deliver similar functionality to the previous sketch:

#include

TextFinder finder(Serial);

Const int NUMBER_OF_FIELDS = 3; // how many comma-separated fields we expect

Int fieldIndex = 0;

// the current field being received

Int values[NUMBER_OF_FIELDS];

// array holding values for all the fields

Void setup()

{

Serial.begin(9600); // Initialize serial port to send and receive at 9600 baud

}

Void loop()

{

For(fieldIndex = 0; fieldIndex < 3; fieldIndex ++)

{

Values[fieldIndex] = finder.getValue(); // get a numeric value

4.5. Sending Binary Data from Arduino

Here, we intend to transmit data in binary format, because we want to pass information with the least number of bytes or because the application you are connecting to only deals with binary data.

This header of this sketch is accompanied by two integer (16-bit) values as binary data. The Values are assembled using the Arduino random function. The Syntax is below;

```
Int intValue; // an integer value (16 bits)
Void setup()
{
Serial.begin(9600);
}
Void loop()
{
Serial.print('H'); // send a header character
// send a random integer
intValue = random(599); // generate a random number between 0 and 599
// send the two bytes that comprise an integer
Serial.print(lowByte(intValue), BYTE); // send the low byte
Serial.print(highByte(intValue), BYTE); // send the high byte
// send another random integer
intValue = random(599); // generate a random number between 0 and 599
// send the two bytes that comprise an integer
Serial.print(lowByte(intValue), BYTE); // send the low byte
Serial.print(highByte(intValue), BYTE); // send the high byte
Delay(1000);
}
```

4.7. Receiving Binary Data from Arduino on a Computer

You may decide to interact with a binary data transmitted from Arduino in a programming language such as Processing. The way you'll do this is contingent upon the programming environment you utilize on your computer. If you don't have a preferable programming tool yet, and you intend to get one that is easy to learn and works well with Arduino, Processing is a perfect choice. Here is the

Processing code to read a byte, extracted from the Processing SimpleRead.

```
Import processing.serial.*;
Serial myPort;          // Create object from Serial class
Short portIndex = 1;    // select the com port, 0 is the first port
Char HEADER = 'H';
Int value1, value2;     // Data received from the serial port
Void setup()
{
  Size(600, 600);
  // Open whatever serial port is connected to Arduino.
  String portName = Serial.list()[portIndex];
  Println(Serial.list());
  Println(" Connecting to -> " + Serial.list()[portIndex]);
  myPort = new Serial(this, portName, 9600);
}
Void draw()
{
  // read the header and two binary *(16 bit) integers:
  If ( myPort.available() >= 5) // If at least 5 bytes are available,
  {
    If( myPort.read() == HEADER) // is this the header
    {
      Value1 = myPort.read();                  // read the least significant byte
      Value1 = myPort.read() * 256 + value1;   // add the most significant byte
      Value2 = myPort.read();                  // read the least significant byte
      Value2 = myPort.read() * 256 + value2;   // add the most significant byte

      Println("Message received: " + value1 + "," + value2);
    }
  }
  Background(255);     // Set background to white
  Fill(0);             // set fill to black
  // draw rectangle with coordinates based on the integers received from Arduino
  Rect(0, 0, value1,value2);
```

4.8. Sending Binary Values from Processing to Arduino

The operation here is to send binary bytes, integers, or long values from Processing to Arduino. For instance, you intend to send a message that entails a message identifier "tag," an index, referring to a particular device interfaced with Arduino, and a 16-bit value.

-Syntax

```
Import processing.serial.*;
Serial myPort;  // Create object from Serial class
Public static final char HEADER = '|';
Public static final char MOUSE  = 'M';
Void setup()
{
  Size(200, 400);
  String portName = Serial.list()[0];
  myPort = new Serial(this, portName, 9600);
}
Void draw(){
}
Void serialEvent(Serial p) {
  // handle incoming serial data
  String inString = myPort.readStringUntil('\n');
  If(inString != null) {
     Println( inString );   // echo text string from Arduino
  }
}
```

Upon clicking the mouse in the Processing window, sendMessage will be introduced into the vertical position of the mouse in the window when clicked and value equal to the horizontal position. The size of the window is set to 200,400, so index would fit into a single byte and value would fit into two bytes:

```
Void mousePressed() {
    Int index = mouseY;
    Int value = mouseX;
    sendMessage(MOUSE, index, value);
}
```

sendMessage sends a header, tag, and index as single bytes. It sends the value as two bytes, with the most significant byte first:

```
void sendMessage(char tag, int index, int value){
    // send the given index and value to the serial port
    myPort.write(HEADER);
    myPort.write(tag);
    myPort.write(index);
    char c = (char)(value / 256); // msb
    myPort.writeC;
    c = (char)(value & 0xff); // lsb
    myPort.write.C;
}
```

However, it can still be used as an integer. The lines that follow change the two bytes to an integer. Serial.read() * 256; restores the most significant byte to its original value. Compare this to Processing code that sent the two bytes with values;

```
Int val = Serial.read() * 256;

Val = val + Serial.read();

Serial.print("Received mouse msg, index = ");

Serial.print(index);

Serial.print(", value ");

Serial.println(val);

}

Else

{

If the code gets here, the tag was not recognized. This helps you to ignore data that may be inc
or corrupted:

Serial.print("got message with unknown tag ");

Serial.println(tag);

}

}

}

}
```

4.9. Sending the Value of Multiple Arduino Pins

You intend to interact by sending clusters of binary bytes, integers, or long values from Arduino. An example is if you decide to transmit the values of the digital and analog pins to Processing.

This recipe transmits a header, and afterwards an integer containing the bit values of digital pins 2 to 13. After which it sends six integers that contain the values of analog pins 0 through 5.

- Syntax

```
Const char HEADER = 'H';   // a single character header to indicate the start
Of a message
// these are the values that will be sent in binary format
Void setup()
{
  Serial.begin(9600);
  For(int i=2; I <= 13; i++)
  {
    pinMode(I, INPUT);      // set pins 2 through 13 to inputs
    digitalWrite(I, HIGH);  // turn on pull-ups
  }
}
Void loop()
{
  Serial.print(HEADER,BYTE); // send the header
  // put the bit values of the pins into an integer
  Int values = 0;
  Int bit = 0;
  For(int i=2; I <= 13; i++)
  {
    bitWrite(values, bit, digitalRead(i)); // set the bit to 0 or 1 depending
                    // on value of the given pin
    Bit = bit + 1;          // increment to the next bit
  }
```

```
sendBinary(values); // send the integer

for(int i=0; I < 6; i++)
{
  Values = analogRead(i);
  sendBinary(values); // send the integer
}
Delay(1000); //send every second
}


// function to send the given integer value to the serial port
Void sendBinary( int value)
{
  // send the two bytes that comprise an integer
  Serial.print(lowByte(value), BYTE);  // send the low byte
  Serial.print(highByte(value), BYTE); // send the high byte
}
```

4.10. Logging Arduino Data to a File on Your Computer

This activity makes them make a file bearing data that was sent over the sequential port from Arduino. A commonplace model is in the event that you need to store the estimations of the advanced and simple pins at intermittent spans to a log record. The Processing sketch that logs the record is created comparative with the Processing sketch additionally expressed there. This Processing sketch builds up a file with the refreshed date and time as the filename, in a similar registry as the

Processing sketch. Approaching messages from Arduino are remembered for the record. Press any key to save the file and close the program.

- **Syntax**

```
Import processing.serial.*;

PrintWriter output;

DateFormat fnameFormat= new SimpleDateFormat("yyMMdd_HHmm");

DateFormat timeFormat = new SimpleDateFormat("hh:mm:ss");

String fileName;

Serial myPort;     // Create object from Serial class

Short portIndex = 0;  // select the com port, 0 is the first port

Char HEADER = 'H';

Void setup()
{
  Size(200, 200);
  // Open whatever serial port is connected to Arduino.
  String portName = Serial.list()[portIndex];
  Println(Serial.list());
  Println(" Connecting to -> " + Serial.list()[portIndex]);
  myPort = new Serial(this, portName, 9600);
  Date now = new Date();
  fileName = fnameFormat.format(now);
  output = createWriter(fileName + ".txt"); // save the file in the sketch folder
}
```

```
Void draw()
{
  Int val;
  String time;
  If ( myPort.available() >= 15) // wait for the entire message to arrive
  {
    If( myPort.read() == HEADER) // is this the header
    {
      String timeString = timeFormat.format(new Date());
      Println("Message received at " + timeString);
      Output.println(timeString);
      // header found
      // get the integer containing the bit values
      Val = readArduinoInt();
      // print the value of each bit
      For(int pin=2, bit=1; pin <= 13; pin++){
        Print("digital pin " + pin + " = " );
        Output.print("digital pin " + pin + " = " );
        Int isSet = (val & bit);
        If( isSet == 0){
          Println("0");
          Output.println("0");
        }
```

```
Import processing.serial.*;
Serial myPort;        // Create object from Serial class
Short portIndex = 1;  // select the com port, 0 is the first port
Char HEADER = 'H';
Int value1, value2;   // Data received from the serial port
Void setup()
{
  Size(600, 600);
  // Open whatever serial port is connected to Arduino.
  String portName = Serial.list()[portIndex];
  Println(Serial.list());
  Println(" Connecting to -> " + Serial.list()[portIndex]);
  myPort = new Serial(this, portName, 9600);
}
Void draw()
{
  // read the header and two binary *(16 bit) integers:
  If ( myPort.available() >= 5)  // If at least 5 bytes are available,
  {
    If( myPort.read() == HEADER) // is this the header
    {
      Value1 = myPort.read();          // read the least significant byte
      Value1 = myPort.read() * 256 + value1; // add the most significant byte
      Value2 = myPort.read();          // read the least significant byte
      Value2 = myPort.read() * 256 + value2; // add the most significant byte

Int readArduinoInt()
{
  Int val;    // Data received from the serial port
  Val = myPort.read();        // read the least significant byte
  Val = myPort.read() * 256 + val; // add the most significant byte
  Return val;
}
```

Note that you have to set portIndex to the serial port attached to Arduino.

4.11. Sending Data to Two Serial Devices at the Same Time

What we are confronted with here is that we need to communicate information to a sequential gadget, for example, a sequential LCD, yet at first, you have been utilizing the underlying sequential port to connect with your PC. How would we approach this? This isn't an issue on the Arduino mega as it has four equipment sequential ports; simply create two sequential articles, in which one is utilized for the LCD, and the other for the PC

Void setup() {

// initialize two serial ports on a megaL

Serial.begin(9600);

Serial1.begin(9600);

}

If your Arduino board only has one hardware serial port, it is necessary that you create an emulated or "soft" serial port.

Pick two accessible computerized pins, one for communication, and the other for receiving, and join your sequential gadget to them. It is a lot quicker to utilize the equipment sequential port for connecting with the PC because a USB connector is on the board. Interface the gadget's communicate line to the get pin and the get line to the send pin. In the picture underneath, pin 2 has been picked as the get pin and pin 3 as the communicate pin

Make a NewSoftSerial object in your portray and educate it about the pins to pick as your copied sequential port. In this model, we're making an article named serial_lcd, which we educate to utilize pins 2 and 3:

```
#include <NewSoftSerial.h>

Const int rxpin = 2;      // pin used to receive from LCD
Const int txpin = 3;      // pin used to send to LCD
NewSoftSerial serial_lcd(rxpin, txpin); // new serial port on pins 2 and 3

Void setup()
{
  Serial.begin(9600); // 9600 baud for the built-in serial port
  Serial_lcd.begin(9600); //initialize the software serial port also for 9600
}

Int number = 0;

Void loop()
{
  Serial_lcd.print("The number is "); // send text to the LCD
  Serial_lcd.println(number);   // print the number on the LCD
  Serial.print("The number is ");
  Serial.println(number);       // print the number on the PC console
  Delay(500); // delay half second between numbers
  Number++;  // to the next number
}
```

This sketch opines that a serial LCD has been interfaced to pins 2 and 3 as shown in Figure 4-5, and that a serial console is attached to the built-in port. The loop will repeatedly display the same message on each:

The number is 0

The number is 1

...

4.12 Using Arduino with Raspberry Pi 4

Here are some basic ways to connect these two gadgets;

- Purchase an add-on board like the Gertboard, which is built with an Arduino compatible IC.

- Plug an Arduino board into the USB port of the Raspberry Pi. It is the easiest method.

- Create the connection using a USB to Serial adapter, which has a less sophisticated Arduino like a Pro Mini or a self-made Shrimp.

- Utilize the Serial Pins on the Raspberry Pi to interface with an Arduino. This is the cheapest method and less stressful.

Speaking of the hardware, you connect the 3.3V/GND/TX/RX pins on the Raspberry Pi through a level converter to 5V/GND/RX/TX pins on an Arduino. Conversely, you can purchase a 3.3V Arduino and get by the need for a level converter.

Furthermore, there are some software adjustments that should be a made, such as RPi software changes that involves stating out this line in /etc/inittab with a #

T0:23:respawn:/sbin/getty -L ttyAMA0 115200 vt100

And removing the following parts of the one line in /boot/cmdline.txt

Console=ttyAMA0,115200 kgdboc=ttyAMA0,115200

Additionally, you need to create a link to the serial port so that the Arduino IDE can identify it. The code below does that.

Sudo ln -s /dev/ttyAMA0 /dev/ttyUSB9

This last procedure must be carried out after every reboot. Nevertheless, the Arduino can't be programmed from the IDE running on the Raspberry Pi with these changes. We have to toggle the reset pin on the Arduino to begin programming.

Connect Pin 11 (GPIO 17) of the RPi to the DTR Pin on the Arduino Pro Mini by using the level converter. Run thesevcommands to download and configure avrdude-rpi:

You have to run the Arduino IDE as root from this point i.e. in LXTerminal:

Sudo Arduino

Then, reboot the Raspberry Pi to activate the earlier procedures. You will then be able to program whatever you like on to the Arduino through the IDE running on the Raspberry Pi.

CHAPTER FIVE

5.1 Using a Switch with the Arduino

Using a switch with the Arduino allows us to control certain functions of the Arduino by using a push button. For instance, with the switch, we can switch an LED on or off when we press the push button.

The necessary materials needed to do this are:

- A push button switch
- An Arduino board
- A resistor of any value
- 2 Jumper cables (breadboard jumpers)
- 1 Breadboard
- A Light Emitting Diode

• Place the switch in the breadboard and insert an LED with the longer leg into pin 13 and shorter leg to the Gnd of the Arduino.

- Next, place one end of the resistor in +5 V and then connect the other end with one of the terminals of the switch. Attach the equivalent terminal to Gnd of the Arduino. The terminal that is located on the same side of the first one is the equivalent terminal.

- Afterwards, connect the terminal with the resistor to pin 2 on the Arduino and enter the program:

Int d=2; // to store on or off value

Void setup()

{pinMode(2,INPUT);

pinMode(13,OUTPUT);

}

Void loop()

{

D=digitalRead(2);

If(d==0)

{digitalWrite(13,HIGH);}

Else

{digitalWrite(13,LOW);}

}

That is all! Just press the switch and the LED will illuminate.

5.2 Using a Switch without external resistors

The resistor seems necessary for proper functioning of a button, and every user might prefer using it. Notwithstanding, a little adjustment is included in each Arduino pin. Each pin is initially built with a pull-up resistor that we can activate with just a slight change in our code. The necessary components needed for this operation are:

An Arduino board connected to a computer via USB and A push button.

- **How to Connect Them**

Just insert a terminal on the button in the Arduino GND and attach a digital pin to the other button terminal. In the image below, the pin 12 is used. Most buttons are built with the standard through-hole terminals in which we can directly insert the pins into the terminals on the Arduino

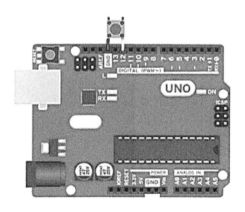
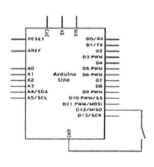

- **The code**

These codes will read when you press the button and will control the built-in LED:

```
// Declare the pins for the Button and the LED<br>int buttonPin = 12;
Int LED = 13;
Void setup() {
    // Define pin #12 as input and activate the internal pull-up resistor
    pinMode(buttonPin, INPUT_PULLUP);
    // Define pin #13 as output, for the LED
    pinMode(LED, OUTPUT);
}

Void loop(){
    // Read the value of the input. It can either be 1 or 0
    Int buttonValue = digitalRead(buttonPin);
    If (buttonValue == LOW){
        // If button pushed, turn LED on
        digitalWrite(LED,HIGH);
    } else {
        // Otherwise, turn the LED off
        digitalWrite(LED, LOW);
    }
}
```

Perhaps the button is inserted into a different pin, change the buttonPin value to the value of the pin that is being used.

5.3. Reliably Detecting when a switch is pressed (Debounce method)

The debounce technique verifies whether the Arduino gets a similar reading from the switch after a delay that should be long enough for the switch contacts to quit bouncing. You may require longer spans for "bouncer" switches (a few switches can require as much as 50 ms or more). The function operates by consistently checking the condition of the switch for the same number of milliseconds as indicated in the debounce time. If the switch stays stable for this time, the condition of the switch will be returned (if pressed, it is true, it's false if it's not pressed). On the off chance that the switch state changes during the debounce period, the counter is reset so it begins again until the switch state doesn't change inside the debounce time. This debounce() function is effective for most switches, yet you should ascertain that the pins being used are in input mode.

The needed components are:

- An Arduino board
- Breadboard
- Dupont Wires
- A Light Emitting Diode

- Tactile Push Button
- A Resistor(220 ohms)
- A 10k resistor.

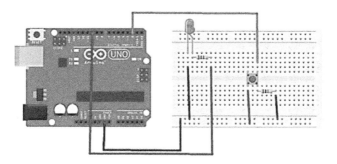

The Code

```
Const int debounceDelay = 10; // milliseconds to wait until stable

// debounce returns true if the switch in the given pin is closed and stable

Boolean debounce(int pin)
{
Boolean state;
Boolean previousState;
previousState = digitalRead(pin); // store switch state

for(int counter=0; counter < debounceDelay; counter++)
{
Delay(1); // wait for 1 millisecond
State = digitalRead(pin); // read the pin
If( state != previousState)
{
Counter = 0; // reset the counter if the state changes
previousState = state; // and save the current state
}
}
// here when the switch state has been stable longer than the debounce period
```

```
Return state;
}
Void setup()
{
pinMode(inputPin, INPUT);
pinMode(ledPin, OUTPUT);
}
Void loop()
{
If(debounce(inputPin))
{
digitalWrite(ledPin, HIGH);
}
}
```

5.4. Determining how long a switch is pressed

Arduino can only determine the state of your button whether it is pressed or not.

You could employ a timer variable to note the exact time you pressed or released the button, with that you can use the difference between both variables to calculate how long it was held.

- The code

The code line is on the next page.

```
Const int buttonPin = 2;
Int buttonState = 0;    // current state of the button
Int lastButtonState = 0; // initial state
Int startPressed = 0;   // the moment you pressed the button
Int endPressed = 0;     // the moment the button was released
Int holdTime = 0;       // how long you held the button
Int idleTime = 0;       // how long the button was idle


Void setup() {
  pinMode(buttonPin, INPUT); // begins button pin as an input
  Serial.begin(9600);     // initiates serial communication
}
Void loop() {
  buttonState = digitalRead(buttonPin); // read the button input
  if (buttonState != lastButtonState) { // changes the state of the button
    updateState();
  }
  lastButtonState = buttonState;    // saves state for next loop
}
Void updateState() {
 If (buttonState == HIGH) {
    startPressed = millis();
    idleTime = startPressed - endPressed;
    if (idleTime >= 500 && idleTime < 1000) {

       Serial.println("Button was idle for half a second");
    }


    If (idleTime >= 1000) {
       Serial.println
    }
 // the button has been released
 } else {
    endPressed = millis();
    holdTime = endPressed - startPressed;
    if (holdTime >= 500 && holdTime < 1000) {
```

        Serial.println

    }

    If (holdTime >= 1000) {

        Serial.println

    }

 }

}

5.5. Reading a Keypad

Keypads are utilized in a vast majority of devices, including cell phones, microwaves, ovens, door locks, etc. A large number of electronic devices use them for user input. Therefore, being able to connect a keypad to a microcontroller such as an Arduino is worth it in case we decide to build various types of commercial products.

After setting everything and all is connected properly and programmed, when a key is pressed, you will see it on the Serial Monitor on your computer.

The components needed are:

- Arduino
- 4x4 Matrix Keypad
- 8 male to male pin header

The type of keypad we will use for this project is a matrix keypad. The matrix keypad operates on an encoding pattern that permits it to have significantly lesser output pins than there are keys. The matrix keypad has 16 keys (0-9, A-D, *, #), however it has only 8 output pins. Because it has lesser pins, it requires lesser connections that need to be set up for the keypad to function. In this manner, the matrix is more efficient.

While joining the pins to the Arduino board, they should be associated with the advanced yield pins, D9-D2. The principal pin of the keypad must be joined to D9, the second pin to D8, the third pin to D7, the fourth pin to D6, the fifth pin to D5, the 6th pin to D4, the seventh pin to D3, and the eighth pin to D2.

These are the connections in a table:

Keypad Pin	Connects to Arduino Pin...
1	D9
2	D8
3	D7
4	D6
5	D5
6	D4
7	D3
8	D2

All the connections are displayed above.

Having settled the physical connection part, let's come to the code. For effective running of this program, you have to import the Keypad library and afterwards, you can enter it into your program. After inputting it into your program, you should notice the line #include <Keypad.h>. If this line isn't displayed, it denotes that the Keypad library has not been successfully inputted into your code, therefore it won't work.

Subsequent to downloading the keypad, change the name of the folder to something different from Keypad. If the folder and the file you are importing have the same name, it won't work. The code is written below;

```
#include <Keypad.h>
Const byte numRows= 4; //the number of rows on the keypad
Const byte numCols= 4; // the number of columns on the keypad
//keymap informs us that the key that is pressed according to the row and columns just as it is shown on the keypad
Char keymap[numRows][numCols]=
{
{'1', '2', '3', 'A'},
{'4', '5', '6', 'B'},
{'7', '8', '9', 'C'},
{'*', '0', '#', 'D'}
};
// indicates the keypad connections to the arduino terminals
Byte rowPins[numRows] = {9,8,7,6}; //Rows 0 to 3
Byte colPins[numCols]= {5,4,3,2}; //Columns 0 to 3

//initializes an instance of the Keypad class
Keypad myKeypad= Keypad(makeKeymap(keymap), rowPins, colPins, numRows, numCols)
Void setup()
{
Serial.begin(9600);
}
//if count=17, then count is reset back to 0 (this implies no key is pressed during the entire keypad scan process
Void loop()
{
Char keypressed = myKeypad.getKey();
```

```
If (keypressed != NO_KEY)

{

Serial.print(keypressed);

}

}
```

After inputting these codes, anytime a key is pressed on the keypad, we should see it on the serial monitor of the Arduino software as soon as the code is compiled and uploaded to the Arduino board.

5.6. Reading analog values

The first step is to wire up the Arduino to read an analog value such as the voltage, determined by the resistance generated by the photoresistor. You can just connect the wires as it is shown in the diagram.

The components you will need are:

- Arduino board
- Photoresistor (5528)
- 10k-Ohm ¼ Watt resistor (Brown)

The A0-A5 pins on the Arduino provide you with a channel to interact with analog sensors, such as photoresistors, knobs (potentiometers), and temperature sensors.

The wiring instructions are:

- Insert a photoresistor into the breadboard just like it is in the diagram
- Connect a 10k-Ohm resistor from one side of the photoresistor across several rows.
- Connect the wires, the colours are indicated too.

Writing the code

```
Void setup() {
    // put your setup code here, to run once:
}
Void loop() {
    // put your main code here, to run repeatedly:
}
```

Before setting it up, you will declare a variable for the analog pin that is connected to the photoresistor

```
    //Photoresistor Pin
    Int analogPin = 0;
    Void setup() {
        // put your setup code here, to run once:
    }
}
```

We intend to write the voltage value emanating from the photoresistor to the serial monitor. To do that, the serial monitor will be initiated using the Serial.begin method and pass in the baud rate (bits per second).

```
Void setup() {
  // put your setup code here, to run once:
  Serial.begin(9600);
}
```

Afterwards, you will enter the code to read the raw data coming in on A0 (be a value between 0 and 1023 which is 1024 steps or units) and convert it to a voltage reading (0.0V to 5.0V).

```
Void loop() {
  // put your main code here, to run repeatedly:
  // read the raw data coming in on analog pin 0:
  Int lightLevel = analogRead(analogPin);
  // Convert the raw data value (0 – 1023) to voltage (0.0V – 5.0V):
  Float voltage = lightLevel * (5.0 / 1024.0);
  // write the voltage value to the serial monitor:
  Serial.println(voltage);
}
```

After the sketch has compiled and you have uploaded it to the Arduino, press the magnifying glass icon in the Arduino IDE. This will open the Serial Monitor, after which the analog values will be displayed.

5.7. Changing the Range of Values

The operation that converts the range of certain values into another range of values is the map function. A typical use is to read an analogue input, which is usually 10 bits long, therefore the values range from 0 to 1023, and change the output to a byte so the output ranges from 0 to 255. Additionally, the map function can convert positive ranges to negative ranges.

Afterwards, input this code;

Val = map(adc_val, 0, 1023, 0, 255);

The map() function looks easy to use until you dive deeper. An even distribution is what you expect but will it really give you that?

Let's compose a sketch to see what it does.

```
Void setup() {
  Serial.begin(9600);
  For (int adc = 0; adc < 1024; adc++) {
    Int mapped = map(adc, 0, 1023, 0, 255);
    Serial.print(adc);
    Serial.print(',');
```

```
    Serial.println(mapped);

  }

}

Void loop() {

}
```

First part of the serial output	Final part of the serial output
0,0	1007,251
1,0	1008,251
2,0	1009,251
3,0	1010,251
4,0	1011,252
5,1	1012,252
6,1	1013,252
7,1	1014,252
8,1	1015,253
9,2	1016,253
10,2	1017,253
11,2	1018,253
12,2	1019,254
13,3	1020,254
14,3	1021,254
15,3	1022,254
	1023,255

The table above contains the different range of values, the value to the left of the comma is the adc value while to the right is the mapped output value.

You will observe that that input is mapped to output ranges in blocks of 4 (1 output value for a range of 4 input values). This is understandable since $1024/4 = 256$. However, the final output value has only 1 input for one output i.e. 1023 results in 255, while 1019~1022 results in 254 as output. Your aim is to obtain an even spread of values across the entire range.

To arrive here, a few of the other outputs must have had 5 values as inputs.

5.9. Measuring voltages greater than 5V

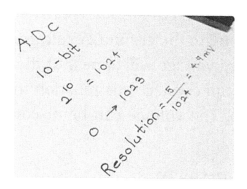

A microcontroller like an Arduino doesn't understand analog voltage directly. That is the importance of an Analog to Digital Converter or ADC in short. The core of the Arduino Uno, Atmega328, has 6 channel (marked as A0 to A5), 10-bit ADC. This implies that it will map input voltages from 0 to 5V into integer values from 0 to (2^10-1) i.e. equal to 1023 which gives a resolution of 4.9mV per unit. 0 will correspond to 0V, 1 to 4.9mv, 2 to 9.8mV and so on till 1023.

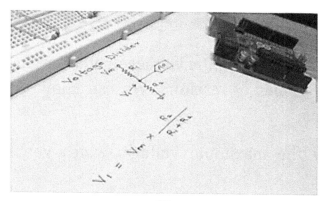

However, the issue comes up when the voltage to be measured exceeds 5 volts. This can be settled by utilizing a voltage divider circuit, which entails 2 resistors connected in series as shown. A part of this series connection is attached to the voltage to be measured (Vm) and the other end to the ground. A voltage (V1) equivalent to the measured voltage will appear at the junction of two resistors. You can connect this junction to the analog pin of the Arduino. The voltage can be discovered using this formula.

V1 = Vm * (R2/(R1+R2))

The Arduino will then measure the voltage V1.

Building the Voltage Divider.

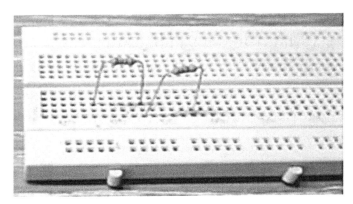

- Next task is to set up the voltage divider; we first need to know the values of resistors. We can know that though these steps:

- Identify the maximum voltage, which you intend to measure.

- Set a suitable and standard value for R1 in the kilo-ohm range.

Using formula, calculate R2.

If the value of R2 isn't precisely the standard value, change R1 and repeat the above steps.

Since Arduino can handle a maximum of 5V, V1 = 5V.

Let's assume values and set the maximum voltage (Vm) to be measured by 12V and R1 = 47 kilo-ohms. Then using the formula R2 comes out to be equal to 33k.

Now, build a voltage divider circuit integrating these resistors.

Following this set up, we are left with a maximum and minimum limit. For Vm = 12V we get V1 = 5V and for Vm = 0V we get V1 = 0V. That is, for 0 to 12V at Vm, we will have an equivalent voltage from 0 to 5V at V1 which can then be inputted into the Arduino as before.

By making slight alterations in the code, we can now measure 0 to 12V. Analog value is read as before. Then, with the aid of the same formula mentioned previously, the voltage between 0 and 12V is measured.

Value = analogRead(A0);

Voltage = value * (5.0/1023) * ((R1 + R2)/R2);

The usual Voltage Sensor Modules are simply a voltage divider circuit. These are rated for 0 to 25V with 30 kilo ohm and 7.5 kilo-ohm resistors.

CHAPTER SIX

6.1 Detecting Movement

In this project, we'll go through how to use a Passive Infrared (PIR) sensor to detect any movement that occurs in your place of residence. A PIR is used in residential and commercial buildings and enables you to detect movement with the Arduino. It does this by registering the latent heat emitted by a person or object.

A PIR sensor is built with three wires: red, brown, and black. The red wire is the power source and should be connected to 5V. While the black wire is the signal wire and not the ground. The brown one should be wired to ground and black to pin 2.

The essential components needed are:

- An Arduino Uno
- A breadboard
- An SE-10 PIR Motion Sensor
- A 10k ohm resistor
- Jump wires

A PIR Sensor

Layout the circuit as in the layout and circuit diagrams.

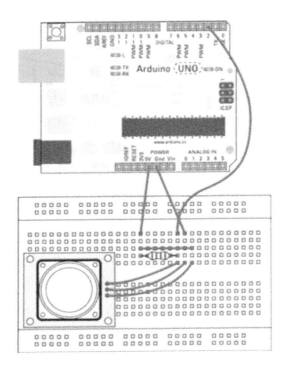

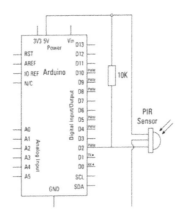

Set up the circuit and select File→Examples→01.Basics→ DigitalReadSerial from the Arduino menu to load the sketch.

This sketch is equivalent to a pushbutton but follows the same method

```
/*
DigitalReadSerial
// digital pin 2 has a pushbutton attached to it. Assign a name to it:
Int pushButton = 2;
// the setup routine runs once when you reset it:
Void setup() {
// begins the serial communication at 9600 bits per second:
Serial.begin(9600);
// make the pushbutton's pin an input:
pinMode(pushButton, INPUT);
}
// the loop routine runs over and over again forever:
Void loop() {
// read the input pin:
Int buttonState = digitalRead(pushButton);
Serial.println(buttonState);
Delay(1);  // delay in between reads for stability
}
```

Click on the compile button to inspect the code. This highlights and points out any error or bug. After the compiling process, click Upload to transfer the sketch to your board. After uploading, place the PIR sensor to a surface that is free of movement and open the serial monitor.

The sketch will reset when you open the serial monitor, and the sensor gets itself ready in the first 1 to 2 seconds. When it detects a movement, the buttonState value changes from 1 (no movement) to 0 (movement). On the other hand, if no changes occur check your wiring.

6.2. Detecting Light

Needed components;

- An Arduino Board
- Breadboard
- LDR
- LED (Color doesn't matter)
- 10k ohm Resistor
- 220 ohm Resistor
- 5 Jumper Wires

Begin by connecting the LED. The shorter leg of the LED, which is the cathode (-), is to be attached to the Ground of the Arduino (GND). The longer leg(anode, +) should be attached to one terminal of the 220-ohm resistor, with the other end being inserted into Digital Pin 13 of the

Arduino. The LED is now in place. Now we move to the LDR. One of its terminals connects to 5V and the opposite terminal to GND via the 10k resistor. Lastly, connect the same row that is directed to the ground to Analog Pin A0 of the Arduino. Create this connection after the resistor.

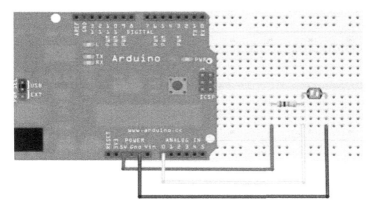

Next thing is to connect the Arduino board to your PC and upload this code;

Const int ledPin = 13;

Const int ldrPin = A0;

Void setup() {

```
  pinMode(ledPin, OUTPUT);

  pinMode(ldrPin, INPUT);

}
Void loop() {

  Int ldrStatus = analogRead(ldrPin);

  If (ldrStatus <=300) {

    digitalWrite(ledPin, HIGH);

    delay(1000);

  }
  Else {

    digitalWrite(ledPin, LOW);

  }
  Delay(100);

}
```

Compile it and run the code, your Arduino will conveniently detect changes in light levels in the LED.

6.3. Measuring Distance Precisely

Precise distance measurement on Arduino can be done via two sensors, which are the infrared proximity sensor and the ultrasonic range finder. They operate in a similar fashion and achieve almost the same thing, but it's vital to choose the appropriate sensor for the environment you're in. We'll focus on using the ranger. The range finder needs to be soldered to your header pins or on lengths of wire if it will be used on a breadboard. We'll use the pulse width method to connect the range finder and convert that to distance.

The components needed include:

- An Arduino Uno
- An LV-EZ0 Ultrasonic Range Finder
- Jump wires

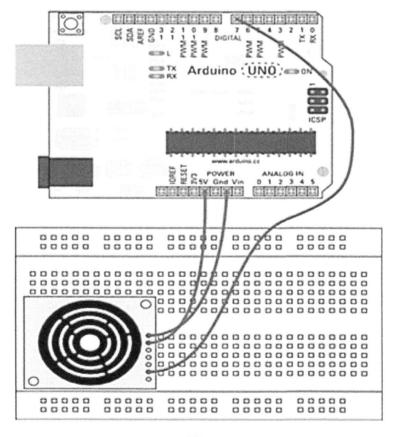

Set up the circuit from the layout and circuit diagrams. The connections for the rangefinder are indicated beneath the PCB. The 5V and GND connections provide power for the sensor and should be attached to the 5V and GND supplies on your Arduino. The PW connection represents the pulse width signal that will be read by pin 7 on your Arduino. Ensure that your distance sensor is positioned firmly to a support in the direction that you intend to measure.

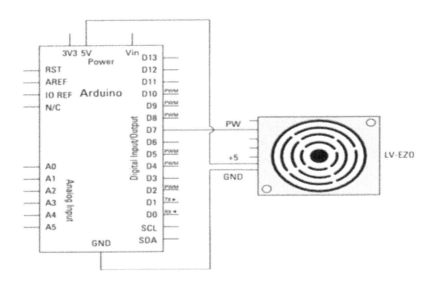

Create a new sketch, enter these codes into it, and save it with.

```
Const int pwPin = 7;
Long pulse, inches, cm;
Void setup() {
  Serial.begin(9600);
}
Void loop() {
  pinMode(pwPin, INPUT);
  //Pulse Width representation with a scale factor of 147 us per Inch.
  Pulse = pulseIn(pwPin, HIGH);
  //147uS per inch
  Inches = pulse/147;
  //change inches to centimetres
  Cm = inches * 2.54;
  Serial.print(inches);
  Serial.print("in, ");
  Serial.print(cm);
  Serial.print("cm");
  Serial.println();
  Delay(500);
}
```

Compile the code, and then upload it to your board. After uploading, you should see the distance measured in inches and centimetres on your serial monitor. If the value isn't stable, try with an entity which has a larger surface.

You can even check the result with a measuring tape and make corrections to the code if you notice irregularities.

6.4. Detecting vibration

In this task, we will connect Arduino with a Vibration sensor and Light Emitting Diode. When the Arduino doesn't detect any vibration, the vibration sensor output is 0 which implies low voltage, otherwise, it's output is 1 (high voltage). In cases where the Arduino gets 0 (no vibration) from the vibration sensor, it will exhibit the green LED and turn off Red LED. If Arduino gets 1 from the vibration sensor, it will exhibit the Red LED and turn off the green LED.

The components needed include;

- A vibration sensor module
- 3 Jumper wires
- A generic LED
- An Arduino board.

Create a new sketch and input this code;

```
Int vib_pin=7;

Int led_pin=13;

Void setup() {

  pinMode(vib_pin,INPUT);

  pinMode(led_pin,OUTPUT);
```

```
}
Void loop() {
  Int val;
  Val=digitalRead(vib_pin);
  If(val==1)
  {
    digitalWrite(led_pin,HIGH);
    delay(1000);
    digitalWrite(led_pin,LOW);
    delay(1000);
  }
  Else
    digitalWrite(led_pin,LOW);
}
```

Compile it and upload it to your Arduino board.

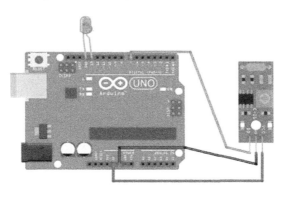

6.5. Detecting Sound

This aspect will help us with how to use the KY-037 sound detection sensor with Arduino. You can measure the changes in the intensity of sound around you through this module.

The needed components are:

- An Arduino board
- ElectroPeak KY-037 Sound Detection Sensor Module
- Bargraph
- 330 Ohm Resistor (10)
- ElectroPeak Male to Female jumper wire

Connect the components and set up a circuit as shown below;

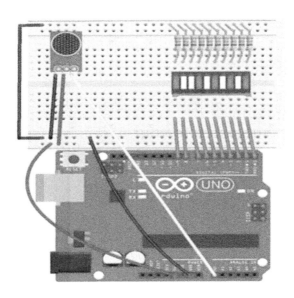

After that, open a new sketch and input this code:

```
Void setup() {
Serial.begin(9600);
}
Void loop() {
Serial.println(analogRead(A0));
Delay(100);
}
```

Then, launch the Serial Monitor window. Now turn the potentiometer to disable the LED on digital output. Jot the number shown in the Serial Monitor immediately after the LED turns off.

Input the codes you jotted before into these codes as a threshold variable and upload it on your board.

```
*/
Int sensor_value = 0;
Int threshold = 540; //Enter your threshold value
Int abs_value = 0;
Int ledCount = 10; //number of Bargraph LEDs
Int bargraph[] = {0, 1, 2, 3, 4, 5, 6, 7, 8, 9}; // Bargraph pins
Void setup() {
Serial.begin(9600); // setup serial
```

```
For (int I = 0; I <= ledCount; i++) // Define bargraph pins OUTPUT
{
pinMode(bargraph[i], OUTPUT);
}
For (int I = 0; I <= 9; i++)
{
digitalWrite(I, LOW);
}
}
Void loop() {
Sensor_value = analogRead(A0);
Abs_value = abs(sensor_value – threshold);
Int ledLevel = map(abs_value, 0, (1024 – threshold), 0, ledCount);
For (int I = 0; I < ledCount; i++) {
// if the array element's index is less than ledLevel,
// turn the pin for this element on:
If (I < ledLevel) {
digitalWrite(bargraph[i], HIGH);
Serial.println(i);
}
```

// turn off all pins higher than the ledLevel:

Else {

digitalWrite(bargraph[i], LOW);

}

}

}

6.6. Measuring Temperature

Items needed;

- An Arduino board
- Lm35 temperature sensor
- A breadboard
- Jumper wires

Temperature measurements can be performed with Arduino if an lm35 temperature sensor is interfaced with it. The lm35 is an analog linear temperature sensor; this implies that the output voltage is equivalent to the temperature. The output voltage increases by 10mv for every 1 degree Celsius rise in temperature.

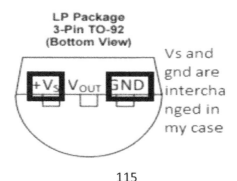

Connect the Vs pin on the lm35 to the 5v pin on the Arduino and attach the ground pin to one of the 2 ground pins on the power rail. Connect the Vout pin to one of the analog pins.

Create a new sketch and input these codes;

Int temppin=0;

Float temp;

Void setup()

{

Serial.begin(9600);

}

Void loop()

```
{
Temp=analogRead(temppin);
Temp=(5.0*temp*1000.0)/(1024*10);
/* 5*temp/1024 is to convert the 10 bit number to a voltage reading.
This is multiplied by 1000 to convert it to millivolt.
We then divide it by 10 because each degree rise results in a 10 millivolt increase.
*/
Serial.println(temp);
Delay(500);
}
```

After that, compile the code and upload it to your Arduino. Launch the serial monitor and observe the temperature, the reading changes every 0.5 seconds.

6.7. Reading RFID (NFC) tags

RFID means radio-frequency identification. In an RFID, electromagnetic fields are used to transfer data over short distances. An RFID device uses tags and readers to operate.

To incorporate an RFID with an Arduino, you'll have to download the RFID library, unzip the library and install it on your Arduino, then you'll restart the Arduino.

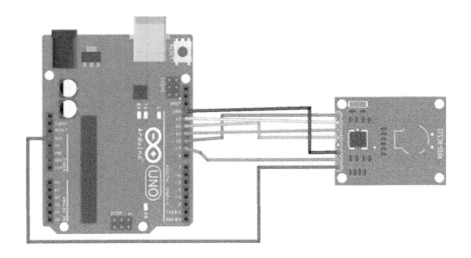

Pin wiring

Pin	Wiring to Arduino Uno
SDA	Digital 10
SCK	Digital 13
MOSI	Digital 11
MISO	Digital 12
IRQ	unconnected
GND	GND
RST	Digital 9
3.3V	3.3V

There's a code already available in your Arduino when you install the RFID library. After setting up the circuit, go to File > Examples > MFRC522 > DumpInfo and upload the code. After uploading, open the serial monitor and you'll see something of this nature;

Approximate the RFID card or the keychain to the reader. Let the reader and the tag closer until it provides you with all the information. This is the information that the card provides you with, including the card UID. The memory is divided into segments and blocks, and information is stored in the memory.

You have 1024 bytes of data storage separated into 16 sectors. Each area is guarded by two different keys, A and B.

Note your UID card and upload this code:

#include <SPI.h>

```
#include <MFRC522.h>

#define SS_PIN 10

#define RST_PIN 9

MFRC522 mfrc522(SS_PIN, RST_PIN);   // Create MFRC522 instance.

Void setup()

{

  Serial.begin(9600);   // Initiate a serial communication

  SPI.begin();      // Initiate  SPI bus

  Mfrc522.PCD_Init();   // Initiate MFRC522

  Serial.println("Approximate your card to the reader...");

  Serial.println();

}

Void loop()

{

  // Look for new cards

  If ( ! mfrc522.PICC_IsNewCardPresent())

  {

    Return;

  }

  // Select a card
```

```
If ( ! mfrc522.PICC_ReadCardSerial())
{
  Return;
}
//Show UID on serial monitor
Serial.print("UID tag :");
   Serial.print(mfrc522.uid.uidByte[i] < 0x10 ? " 0" : " ");
   Serial.print(mfrc522.uid.uidByte[i], HEX);
   Content.concat(String(mfrc522.uid.uidByte[i] < 0x10 ? " 0" : " "));
   Content.concat(String(mfrc522.uid.uidByte[i], HEX));
  }
  Serial.println();
  Serial.print("Message : ");
  Content.toUpperCase();
  If (content.substring(1) == "BD 31 15 2B") //change here the UID of the card/cards that you want to give access
  {
    Serial.println("Authorized access");
    Serial.println();
    Delay(3000);
  }
  Else {
    Serial.println(" Access denied");
    Delay(3000);
  }
}
```

Approximate the card you've picked to give access and it will display this;

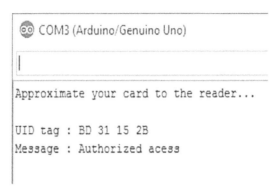

Upload it to your Arduino and that's that.

6.8. Using Mouse Button Control

You can be in charge of the computer's onscreen cursor with an Arduino. This device utilizes five pushbuttons to move the cursor. Four of these buttons are directional (up, down, left, right) and one is for a left mouse click. Arduino exhibits a relative cursor movement. Every time it reads an input, the cursor's position is updated relative to its present position.

The fifth button is in charge of the left-click function from the mouse. When the button is released, the computer will identify this operation.

Components Required;

- A Breadboard
- An Arduino board
- 10k ohm resistor(5)

- Momentary pushbuttons(5)

Set up the circuit as shown in the Image;

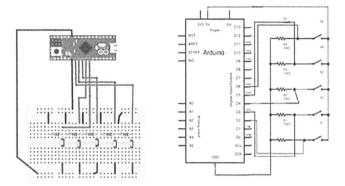

Then launch the Arduino IDE software on your computer, create a new sketch and input these codes;

```
#include "Mouse.h"

// set pin numbers for the five buttons:

Const int upButton = 2;

Const int downButton = 3;

Const int leftButton = 4;

Const int rightButton = 5;

Const int mouseButton = 6;

Int range = 5; // output range of X or Y movement; affects movement speed

Int responseDelay = 10; // response delay of the mouse, in ms


Void setup() {

    // initialize the buttons' inputs:

    pinMode(upButton, INPUT);

    pinMode(downButton, INPUT);
```

```
  pinMode(leftButton, INPUT);

  pinMode(rightButton, INPUT);

  pinMode(mouseButton, INPUT);

  // initialize mouse control:

  Mouse.begin();
}


Void loop() {

  // read the buttons:

  Int upState = digitalRead(upButton);

  Int downState = digitalRead(downButton);

  Int rightState = digitalRead(rightButton);

  Int leftState = digitalRead(leftButton);

  Int clickState = digitalRead(mouseButton);

  // calculate the movement distance based on the button states:

  Int xDistance = (leftState - rightState) * range;

  Int yDistance = (upState - downState) * range;

  // if X or Y is non-zero, move:

  If ((xDistance != 0) || (yDistance != 0)) {

    Mouse.move(xDistance, yDistance, 0);

  }
```

```
// if the mouse button is pressed:

If (clickState == HIGH) {

  // if the mouse is not pressed, press it:

  If (!Mouse.isPressed(MOUSE_LEFT)) {

    Mouse.press(MOUSE_LEFT);

  }

} else {              // else the mouse button is not pressed:

  // if the mouse is pressed, release it:

  If (Mouse.isPressed(MOUSE_LEFT)) {

    Mouse.release(MOUSE_LEFT);

  }

}

// a delay so the mouse does not move too fast:

Delay(responseDelay);
```

Upload the codes to your Arduino and that should be it. Move the pushbutton and you'll notice the changes

6.9. Getting location from a GPS

The materials required are:

- An Arduino board
- NEO-6m GPS module

- LCD display

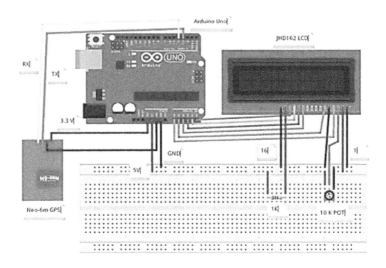

The circuit connection will be in this fashion;

The circuit as follows:

- Neo-6m GPS module == Arduino board
- GND→GND
- TX→Digital pin (D3)
- RX→Digital pin (D4)
- Vcc→3.3 V
- LCD→Arduino * VSS ⍰ GND
- VCC→5V
- VEE→10K Resistor
- RS→A0 (Analog pin)
- R/W→GND
- E→A1
- D4→A2
- D5→A3
- D6→A4

- D7→A5
- LED+→VCC
- LED-→GND

Open the Arduino software on the computer, create a new sketch and input this code;

```
#include <LiquidCrystal.h>

#include <SoftwareSerial.h>

#include <TinyGPS.h>

//long  lat,lon; // develop variable for latitude and longitude object

Float lat ,lon ; // develop variable for latitude and longitude object

SoftwareSerial gpsSerial(3,4);//rx,tx

LiquidCrystal lcd(A0,A1,A2,A3,A4,A5);

TinyGPS gps; // create gps object

Void setup(){

Serial.begin(9600); // connect serial

Serial.println("The GPS Received Signal:");

gpsSerial.begin(9600); // connect gps sensor

lcd.begin(16,2);

}

Void loop(){
```

```
While(gpsSerial.available()){ // check for gps data
If(gps.encode(gpsSerial.read()))// encode gps data
{
Gps.f_get_position(&lat,&lon); // obtain  latitude and longitude
// display position
Lcd.clear();
Lcd.setCursor(1,0);
Lcd.print("GPS Signal");
Lcd.setCursor(1,0);
Lcd.print("LAT:");
Lcd.setCursor(5,0);
Lcd.print(lat);
Serial.print(lat);
Serial.print(" ");
Serial.print(lon);
Serial.print(" ");
Lcd.setCursor(0,1);
Lcd.print(",LON:");
Lcd.setCursor(5,1);
Lcd.print(lon);

}
}
String latitude = String(lat,6);
String longitude = String(lon,6);
Serial.println(latitude+";"+longitude);
Delay(1000);
}
```

After setting up the connection and uploading the code, the GPS module would delay a while, for about 15- 20 minutes, to get a satellite fix, and the LCD screen will display the coordinates.

6.10. Reading Acceleration

Required components:

- ADXL345
- Arduino
- I2C Cable
- I2C Shield for Arduino

The ADXL345 is a miniature ultralow power, 3-axis accelerometer with high-resolution (13-bit) measurement at up to ±16 g. The connection of the ADXL345 sensor module with the Arduino will be illustrated. To read the acceleration values, an Arduino with an I2c adapter is used. This I2C adapter makes the connection to the sensor module easy and more reliable.

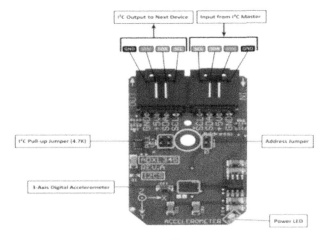

Only four-wire connections are needed Vcc, Gnd, SCL, and SDA pins, and these are connected via the I2C cable.

Open the Arduino program on your computer, create a new sketch and input these codes;

```
Serial.begin(9600);

// Start I2C Transmission

Wire.beginTransmission(Addr);

// Select bandwidth rate register

Wire.write(0x2C);

// Normal mode, Output data rate = 100 Hz

Wire.write(0x0A);

// Stop I2C transmission


Wire.endTransmission();

// Start I2C Transmission
```

```
Wire.beginTransmission(Addr);
// Select power control register
Wire.write(0x2D);
// Auto-sleep disable
Wire.write(0x08);
// Stop I2C transmission
Wire.endTransmission();
// Start I2C Transmission
Wire.beginTransmission(Addr);
// Select data format register
Wire.write(0x31);
// Self test disabled, 4-wire interface, Full resolution, Range = +/-2g
Wire.write(0x08);
// Stop I2C transmission
Wire.endTransmission();
Delay(300);
}
Void loop()
{
Unsigned int data[6];
For(int I = 0; I < 6; i++)
{
// Start I2C Transmission
Wire.beginTransmission(Addr);
// Select data register
```

```
Wire.write((50 + i));
// Stop I2C transmission
Wire.endTransmission();
// Request 1 byte of data
Wire.requestFrom(Addr, 1);
// Read 6 bytes of data
If(Wire.available() == 1)
{
Data[i] = Wire.read();
}
// Convert the data to 10-bits
Int xAccl = (((data[1] & 0x03) * 256) + data[0]);
If(xAccl > 511)
{
xAccl -= 1024;
}
Int yAccl = (((data[3] & 0x03) * 256) + data[2]);
If(yAccl > 511)
{
yAccl -= 1024;
}
Int zAccl = (((data[5] & 0x03) * 256) + data[4]);
If(zAccl > 511)
{
zAccl -= 1024;
```

}

// Output data to serial monitor

Serial.print("Acceleration in X-Axis is : ");

Serial.println(xAccl);

Serial.print("Acceleration in Y-Axis is : ");

Serial.println(yAccl);

Serial.print("Acceleration in Z-Axis is : ");

Serial.println(zAccl);

Delay(300);

}

Upload the codes to your Arduino and run it.

CHAPTER SEVEN

7.0. Visual Output

These projects show the simplest things that can be achieved with the Arduino to see the physical or visual output.

7.1. Connecting and Using LED

Required components;

- An Arduino board
- A light-emitting diode
- 220-ohm resistor
- Circuit Set up

The LED is connected to a digital pin and the number indented on it may differ according to board type. To make things easier, a constant is specified in every board descriptor file. This constant is LED_BUILTIN and it enables you to control the built-in LED easily. Here is the proportionality between the constant and the digital pin;

Digital pin – Constant

D13 – 101

D13 – Due

D1 – Gemma

D13 – Intel Edison

D13 – Intel Galileo Gen2

D13 – Leonardo and Micro

D13 – LilyPad

D13 – LilyPad USB

D13 – MEGA2560

D13 – Mini

D6 – MKR1000

D13 – Nano

D13 – Pro

D13 – Pro Mini

D13 – UNO

D13 – Yún

D13 – Zero

If you want to light up the external LED with your sketch, a circuit which entails the connection of one terminal of the resistor to the digital pin correspondent to the LED_BUILTIN constant. Attach the long leg of the Light Emitting (the positive leg, otherwise known as the anode) to the other terminal of the resistor. Connect the short leg of the LED (the negative leg, called the cathode) to the GND. The diagram below illustrates an Arduino UNO board that has D13 as the LED_BUILTIN value.

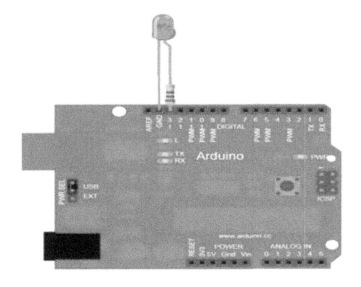

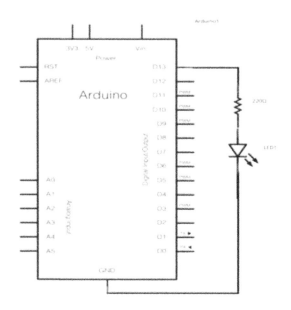

Afterward, interface the Arduino board with your computer, start the Arduino Software (IDE) and input this code;

Void setup()

pinMode(LED_BUILTIN, OUTPUT);

Void loop() {

digitalWrite(LED_BUILTIN, HIGH); // turn the LED on (HIGH is the voltage level)

delay(1000); // wait for a second

digitalWrite(LED_BUILTIN, LOW); // turn the LED off by making the voltage LOW

delay(1000); // hold on for a second

Upload the sketch to your Arduino and run it.

7.2. Adjusting the brightness of an LED

The LED brightness changes with the magnitude of the current flowing through it. If 5V and 3.3V are applied, 5V makes the LED light brighter. In the event that you need the LED to be significantly more lightened, you either "lessen the resistance" or "increment the voltage".

The brilliance can be adjusted through the LED squint. Associate the Light Emitting Diode pin 5 of Arduino, as demonstrated as follows:

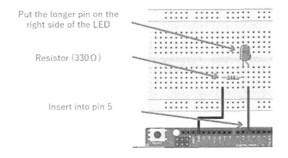

Then, create the sketch and upload the codes, after that, upload it to the Arduino. The LED would blink at an interval of one second.

The code;

Const int LED_PIN = 5;

Const int ON_TIME = 1000;

Const int OFF_TIME = 1000;

Void setup(){

```
  pinMode( LED_PIN, OUTPUT );
}
Void loop(){
  digitalWrite( LED_PIN, HIGH );
  delay( ON_TIME );
  digitalWrite( LED_PIN, LOW );
  delay( OFF_TIME );
}
```

7.3. Adjusting the color of an LED

In this project, we will use an RGB (Red, Green, Blue) LED with an Arduino. The RGB LED can display several colors depending on mixing the 3 basic colors red, green, and blue.

The components needed are;

- Arduino Uno Rev3
- Breadboard 400 point
- Dupont Wires
- RGB LED

Circuit set up:

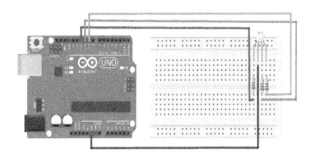

Connect the ground 3 pins which stand for the colors will be connected to the cathode through 220 Ohm resistors to 3 pins on the Arduino that generates the PWM signal. The PWM signal's function is to set various voltage levels to the LEDs to get the intended colors. Create a new sketch on your pc and input these codes, afterwards, upload it on your Arduino.

```
// RGB LED sketch
Int redPin = 11;
Int greenPin = 10;
Int bluePin = 9;
Void setup()
{
  pinMode(redPin, OUTPUT);
  pinMode(greenPin, OUTPUT);
  pinMode(bluePin, OUTPUT);
}
Void loop()
{
  setColor(255, 0, 0); // red
  delay(1000);
  setColor(0, 255, 0); // green
  delay(1000);
  setColor(0, 0, 255); // blue
  delay(1000);
  setColor(255, 255, 0); // yellow
  delay(1000);
  setColor(80, 0, 80); // purple
  delay(1000);
  setColor(0, 255, 255); // aqua
  delay(1000);
}
```

```
Void setColor(int red, int green, int blue)
{
  //use PWM to controlt the brightness of the RGB LED
  analogWrite(redPin, red);
  analogWrite(greenPin, green);
  analogWrite(bluePin);
}
```

7.4. Sequencing Multiple LEDs

An LED sequencer is a typical LED driving circuit that is used in running-light rope displays to flash different lighting patterns. In a sequencer circuit, a controller directs the sequence and timing of the flashing LEDs to exhibit various kinds of lighting patterns. An LED sequencer is built on Arduino UNO.

Components required:

- An Arduino board
- 7 LEDs
- Seven330 Ohms Resistor
- A Breadboard
- Male-to-Male Jumper wires

Circuit connections:

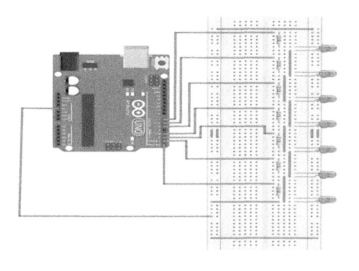

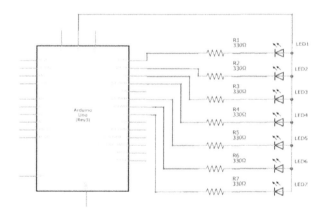

The circuit of an LED sequencer is programmed on a breadboard. LEDs are attached to the Arduino channels to allow them to take the current from the LEDs.

- The cathodes of the LEDs are attached to the digital I/O pins 0, 1, 2, 3, 5, 6, and 7 from the Arduino UNO through the current-limiting series resistor at the individual channel

- Connect the entire LED anodes to the power supplying row of the breadboard.
- Load the sketch for the LED sequencer to the Arduino.
- Utilize the 5V power pin and one of the ground pins from Arduino to supply the 5V DC and then ground to the breadboard.
- Switch on the Arduino by connecting it to a power supply.

Then input the codes for lighting an LED, insert the code such that it encompasses the whole LEDs, for example:

```
Void setup() {

pinMode(0, OUTPUT);

pinMode(1, OUTPUT);

pinMode(2, OUTPUT);

pinMode(3, OUTPUT);

pinMode(5, OUTPUT);

pinMode(6, OUTPUT);

pinMode(7, OUTPUT);

}

digitalWrite(0, HIGH);

delay(200);

digitalWrite(1, HIGH);

delay(200);
```

digitalWrite(2, HIGH);

delay(200);

digitalWrite(3, HIGH);

delay(200);

digitalWrite(5, HIGH);

delay(200);

digitalWrite(6, HIGH);

delay(200);

digitalWrite(7, HIGH);

delay(200);

...

When a logical LOW is fixed as the output from a pin, the current passes through the attached LED and it illuminates. When a logical HIGH is fixed as the output from a pin, the attached LED does not receive the required forward voltage and it will stop illuminating. The LEDs are interfaced to the Arduino in this manner;

LED (from left to right)	Arduino Pin
LED 01	0
LED 02	1
LED 03	2
LED 04	3
LED 05	5
LED 06	6
LED 07	7

7.5. Controlling an LED matrix through multiplexing

Multiplexing is simply a way to split information into little pieces and send it one after the other. With this, you can save countless pins on the Arduino and make your program retain its simplicity. The entirety of the Arduino LED matrix project is based on the principle of multiplexing. Here, the Arduino is connected to the 4017-decade counter IC and transmits the data over two lines. The multiplexed data acquired from the Arduino will be dissected with the goal that it very well may be complete over discrete signs for the LEDs using the 4017 IC. For example, on the off chance that we choose to part a picture into ten pieces, it suggests that we will look over lines of the rows and send information from the Arduino to the sections. All the columns represent the positives of the LEDs and the rows are negatives. Subsequently, if the principal line is associated with the ground and we impart a sign to the primary section, just the main LED in the column will shine. For viable control of the Arduino LED lattice, 4017 IC is the most favored strategy.

The multi decade counter causes us with the multiplexing by essentially checking the lines of the framework and enlightens each line in turn.

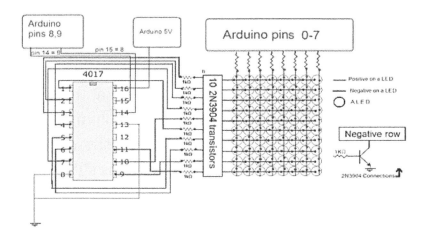

The 4017 is built with 10 output pins, therefore we need 10 resistors and 10 transistors. Next thing is to attach the 1K resistors to the output present on the 4017 and its opposite end to the terminal of the transistor as shown in the circuit. Then, we interface the LED rows to the collectors of the transistors and the emitter to the ground.

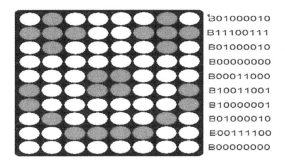

Programming the Arduino LED grid requires controlling the ports. Utilizing ports for programs makes it simpler to work with and it likewise spares space. Port control isn't so difficult, look at the picture for instance. The red specks imply the LEDs, which are ON, and the white as OFF. In the main line of the lattice, to turn on the second and seventh LEDs, we input an order to the Arduino port: B01000010. Here the second and seventh pieces are "1", which thus turns the LED ON. "B" suggests that the port is a yield port. We complete these for all the columns lastly; we will get a smiley face in the lights it is shown in the picture. At that point, unplug the 0 and 1 pins subsequent to transferring the program.

7.6. Using an analog panel meter as display

Required components;

- 2, 5v Analog panel meters

- Arduino board

- DS1307 Real time clock

- Protoboard

- 10K potentiometer

- 2 tactile switches

- 4 10K resistors

- 4 white LEDs (optional)

- USB cable

- 3.5mm cable

The circuit wiring is in a fashion Wiring goes as follows:

- Connect the USB red wire (5v) to VCC

- The USB black wire to GND

- Attach the 3.5mm Audio left channel to 10K resistor to Analog 1

- The 3.5mm Audio right channel to 10K resistor to Analog 2

-Connect the 3.5mm Audio ground to GND

- Potentiometer to Analog 0

- Buttons – Left/Down to Digital 2

    - Right/Up to Digital 3

- DS1307 RTC - SDA to Analog 4

    - SCL to Analog 5

- Left Analog Meter to Digital 5 (PWM)

- Right Analog Meter to Digital 6 (PWM)

Programming this project is somewhat easy, the only tasking aspect is ensuring your meters display accurate values.

```
Const int analogInPin= 0; //Potentiometer wiped connect to pin A0
Const int analogMeterPin= 9; //Analog output pin connected to the meter
Int sensorValue= 0; //value read from the potentiometer
Int outputValue= 0; //value output to the terminal D9 (analog out)
Void setup ()
{
//nothing in setup
}
Void loop()
{
sensorValue= analogRead(analogInPin); //this gives us the analog voltage read from the potentiometer
outputValue= map (sensorValue, 0, 1023, 0, 255); //scale for analog out value
analogWrite (analogMeterPin, outputValue); // write the analog out value
}
```

Upload the code and run it.

Insert it into a USB port and plug the audio cable into a jack splitter. Set the time via the buttons and change the audio sensitivity by using the potentiometer.

CHAPTER EIGHT

8.1. Controlling Rotational position with a servo

Rotational servos are a type of gear-reduced motor with various alterations. The servo rotates in a specific direction when you increase the direction to 90 degree and rotates in the other direction when the angle is less than 90 degree.

The components you will need for this project are:

Arduino – whose function is to transmit electrical pulses to the servo to instruct it on how much to rotate.

- Servo Motor
- 10k Ohm Potentiometer
- Four AA Battery Holder with On Off Switch
- Four AA Batteries
- Solderless Breadboard
- Male-to-Male Jumper Wires

Circuit set up:

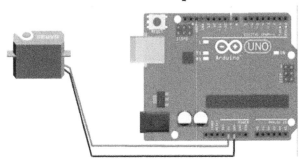

- Attach the red wire of the servo to the 5V pin of the Arduino Uno.

- Insert the black wire of the servo to the GND pin of the Arduino Uno.
- Next thing is to connect the yellow control wire of the servo to Digital Pin 9 of the Arduino Uno. The function of this yellow wire is to receive commands from the Arduino.
- Switch on your Arduino by plugging in the USB cord to your computer.
- Launch the Arduino IDE, and in a new sketch, enter the following code:

```
*/
#include <Servo.h>
Servo myservo;  // create servo object to control a servo
Int pos = 0;    // variable to store the servo position
Void setup() {
  Myservo.attach(9);  // attaches the servo on pin 9 to the servo object
}
Void loop() {
  For (pos = 0; pos <= 180; pos += 1) { // goes from 0 degrees to 180 degrees
    // in steps of 1 degree
    Myservo.write(pos);
    Delay(15);           // waits 15ms for the servo to reach the position
  }
  For (pos = 180; pos >= 0; pos -= 1) { // goes from 180 degrees to 0 degrees
    Myservo.write(pos);
    Delay(15);           // waits 15ms for the servo to reach the position
  }
}
```

- When you're done, upload the code to your board. This code will cause the shaft of the motor sweep back and forth 180 degrees.

8.2. Controlling Servo Rotation with a Potentiometer

In certain instances, we might decide to control the angle a servo rotates without having to always modify the code. This can be achieved by using a potentiometer. The potentiometer is a variable resistor, when you turn the knob, you can regulate the voltage output of the potentiometer. This project requires us to set up software that analyzes the voltage output of the potentiometer. It then converts that number into an angle for the servo.

A potentiometer is built with three terminals;

- **Two outer terminals are used for power:** one outer pin connects to ground and the other connects to positive voltage. Then there's a central control terminal used for voltage output: turning the knob of the potentiometer increases or decreases the resistance, which by implication increases or decreases the voltage output.

Circuit Setup:

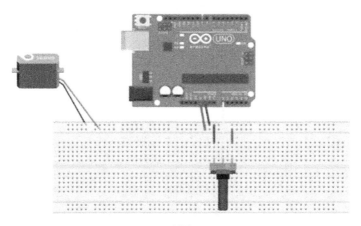

Unplug your Arduino, and insert the 10k Ohm potentiometer into the breadboard. Ensure that each terminal is connected to a separate row in the breadboard. Carry out the connections this way;

- Insert one of the outer terminals to the blue rail of the breadboard.
- Insert into the red (positive) rail of the breadboard the outer terminal
- Connect the central pin of the potentiometer to Analog Input pin A0 of the Arduino.
- Connect the black and red wires of the servo to the blue and red rail of the breadboard, respectively.
- Attach the yellow control wire with pin 9 of the Arduino.
- Insert a wire from the +5V pin of the Arduino to the positive red rail of the breadboard.
- Insert a wire from Ground (GND) of the Arduino to the blue rail of the breadboard.

You can then plug in your Arduino. Launch the Arduino IDE, create a new sketch, enter the following code:

```
#include <Servo.h>
Servo myservo; // create servo object to control a servo
Int potpin = 0; // analog pin used to connect the potentiometer
Int val; // variable to read the value from the analog pin
Void setup() {
  Myservo.attach(9); // attaches the servo on pin 9 to the servo object
}
Void loop() {
  Val = analogRead(potpin);       // reads the value of the potentiometer (value between 0 and 1023)
  Val = map(val, 0, 1023, 0, 180); // scale it to use it with the servo (value between 0 and 180)
  Myservo.write(val);             // sets the servo position according to the scaled value
  Delay(15);
}
```

Transfer the code to your Arduino. Turn the handle on your potentiometer to move the servo, this increment or diminishes the voltage yield, the higher the voltage yield by the potentiometer, the more prominent the servo's point of pivot.

8.3. Controlling a Brushless motor

On the off chance that you expect to decide the method of activity of the Electronic Speed Controller of a Brushless Motor without utilizing a transmitter and recipient, you can do it with an Arduino Microcontroller. This requires utilizing a PWM signal from the Arduino to control the speed of the brushless engine with an ESC.

The needed components are;

- Arduino
- Potentiometer
- Lipo Battery
- ESC
- Brushless Motor

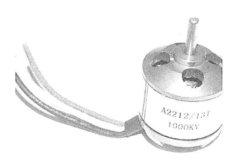

A Brushless motor

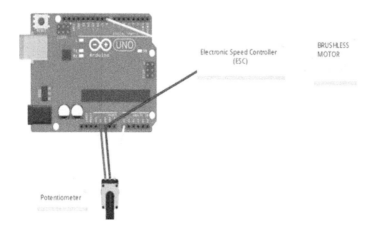

Let's get to the wiring.

Firstly, attach the three terminals of Brushless motor to the three terminals of the ESC. Fix the Motor to a heavy wooden plank or flat surface so that it remains stable at high RPM. Download and Flash the code available at the bottom of the page to the arduino via a USB cable. Connect the signal wire of ESC mostly white or yellow cooler to any PWM pin. You can use more than one pins for controlling many motors.

Associate the Potentiometer to the vcc or 5v pin of the Arduino and the Ground. Interface the third terminal that is the variable pin to the Analog pin A0. You can in any case control the Arduino utilizing the BEC (Battery Eliminator Circuit) accessible in your ESC. If you need to get the BEC, just associate the red thick wire to the Vin Pin of Arduino .It can give 5V. Nevertheless, not all ESC's are given a BEC, in this case you can utilize an outside 5v power gracefully. Force on the Arduino and associate the Lipo battery to your ESC. At that point, bit by bit pivot the Potentiometer Knob to begin and speed up the Motor.

Create a new sketch and input this code;

```
Void setup()
{
Esc.attach(9); //Specify the esc signal pin,Here as D8
Esc.writeMicroseconds(1000); //initialize the signal to 1000
Serial.begin(9600);
}
Void loop()
{
Int val; //Creating a variable val
Val= analogRead(A0); //Read input from analog pin a0 and store in val
Val= map(val, 0, 1023,1000,2000); //mapping val to minimum and maximum(Change if needed)
esc.writeMicroseconds(val); //using val as the signal to esc
}
```

Upload the code to your Arduino, and gradually rotate the knob to adjust the motor's speed.

8.4. Controlling a solenoid

In this project, we want to control a small 5v Solenoid by switching it on and off in intervals of 1 second. The components needed for this project are:

- One Arduino Uno
- A Solderless breadboard
- 5 Jumper Wires
- W 220 Ω Resistor
- Diode
- A Power Transistor

- A 5v Solenoid

Circuit setup:

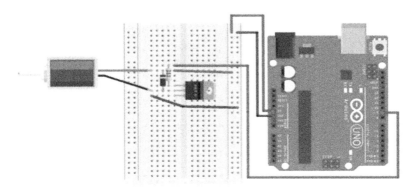

- Insert the 5v Power and Ground from your Arduino to your power and ground rails on your breadboard
- Connect your solenoid to different lines on your breadboard, one to the 5v power from step 2, connect the other one to the collector of the transistor.
- Your Diode should be between the two solenoid cables, this will hinder current discharging back through the circuit when the solenoid coil discharges.
- The power transistor should be inserted on three separate lines of your breadboard, with the flat side facing toward the outside. The collector's leg should be interfaced with the solenoid and diode line.
- The 220-ohm Resistor from the adjacent leg the transistor should be connected to a separate line
- The emitter leg should be inserted into the ground rail. Then, attach the other side of the resistor from step 6 to digital pin 9, which should do it.

Create a new sketch and input this code;

```
Int solenoidPin = 9;        //This is the output pin on the Arduino
Void setup()
{
  pinMode(solenoidPin, OUTPUT);   //Sets that pin as an output
}
Void loop()
{
  digitalWrite(solenoidPin, HIGH);   //Switch Solenoid ON
  delay(1000);                //Wait 1 Second
  digitalWrite(solenoidPin, LOW);    //Switch Solenoid OFF
  delay(1000);
}
```

Upload the code to your Arduino. Now click and keep on clicking, your solenoid should be toggling on and off.

8.5. Driving a brushed motor using a transistor

Driving a brushed motor requires a larger amount of current than an Arduino board can give, therefore you must use a transistor. Transistors are built with limits and maximum specifications, , just ensure those values are enough for your use. The suitable transistor for this tutorial is P2N2222A and is rated at 40V and 200mA.

The components needed for this project are:

- Arduino board
- Breadboard
- 220 Ohm resistor
- Transistor P2N2222A
- Diode 1N4148
- DC Motor

Circuit setup;

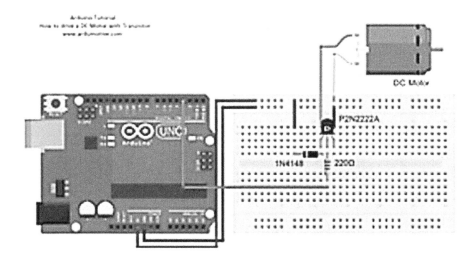

Create a new sketch and input this code;

```
Int motorPin = 3;

Int Speed; //Variable to store Speed, by defaul 0 PWM

Int flag;

Void setup()

{

pinMode(motorPin, OUTPUT); //Set pin 3 as an OUTPUT

        Serial.begin(9600); //Init serial communication
```

```
    //Print a message:
        Serial.println(""); //Blank line
}
Void loop()
{
   If (Serial.available() > 0)
   {
     Speed = Serial.parseInt();
     Flag=0;
   }
//Valid range is from 50 to 255
If (Speed>=50 && Speed<=255){
//Send PWM value with analogWrite to Arduino pin 3 and print a message to serial monitor
analogWrite(motorPin, Speed);
   //Print message only once
   If (flag==0){
        //Print PWM value
Serial.print("Motor spinning with ");
Serial.print(Speed);
Serial.println(" PWM");
Flag=1;
   }
Delay(1000);
}
```

Upload the code to your Arduino and that should do it.

8.6. Controlling a Unipolar stepper

Stepper motors have a unique design and can therefore be controlled to a high degree of accuracy without any feedback mechanisms. The shaft of a stepper, positioned with a set of magnets, is regulated by a series of electromagnetic coils that are charged in a particular sequence, precisely moving its small "steps".

The required hardware are;

- Arduino UNO board
- 28BYJ-48 unipolar stepper motor (with driver board)
- 10k ohm potentiometer
- Pushbutton
- 5V power source
- Bread board
- **Jumper wires**

Circuit setup;

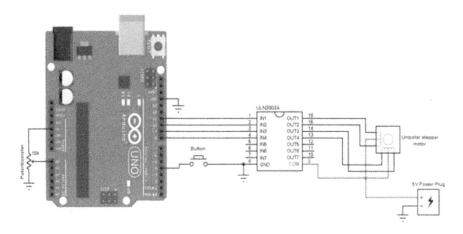

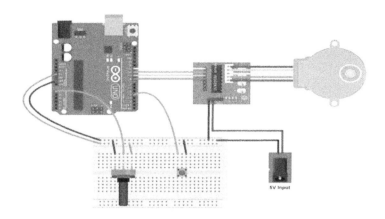

Circuit connection

- Interface the stepper engine to the ULN2003A board, which is given outside force wellspring of 5V. The control lines (IN1, IN2, IN3 and IN4) of this board are interfaced with the Arduino thusly:
- IN1 to Arduino pin 11
- IN2 to Arduino pin 10
- IN3 to Arduino pin 9
- IN4 to Arduino pin 8
- The 10k ohm potentiometer is intended to control the speed of the stepper engine; you will associate its yield pin to Arduino simple pin 0.
- The push button, which is associated with Arduino pin 4, alters the revolution course of the stepper engine.

```
#include <Stepper.h>
// change this to the number of steps on your motor
#define STEPS 32
Stepper stepper(STEPS, 8, 10, 9, 11);
Const int button = 4; // direction control button is connected to Arduino pin 4
Const int pot   = A0; // speed control potentiometer is connected to analog pin 0
```

```
Void setup()
{
  // configure button pin as input with internal pull up enabled
  pinMode(button, INPUT_PULLUP);
}
Int direction_ = 1, speed_ = 0;
Void loop()
{
  If ( digitalRead(button) == 0 )  // if button is pressed
    If ( debounce() )  // debounce button signal
    {
      Direction_ *= -1;  // reverse direction variable
      While ( debounce() ) ;  // wait for button release
    }
  Int val = analogRead(pot);
  // map digital value from [0, 1023] to [2, 500]
  If ( speed_ != map(val, 0, 1023, 2, 500) )
  { // if the speed was changed
    Speed_ = map(val, 0, 1023, 2, 500);
    // set the speed of the motor
    Stepper.setSpeed(speed_);
  }
  // move the stepper motor
  Stepper.step(direction_);
}

Bool debounce()
{
  Byte count = 0;
  For(byte I = 0; I < 5; i++) {
    If (digitalRead(button) == 0)
```

```
  Count++;
  Delay(10);
}
If(count > 2) return 1
```

8.7. Controlling a Bipolar stepper

Hardware required:

- Arduino UNO board
- Bipolar stepper motor
- L293D motor driver chip
- 10k ohm potentiometer
- Pushbutton
- Power source with voltage equal to motor nominal voltage
- Bread board
- Jumper wires

Circuit:

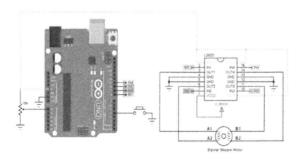

Associate the ground terminals together.

The L293D chip has 16 pins with 4 sources of info (IN1, IN2, IN3 and IN4) and 4 yields (OUT1, OUT2, OUT3 and OUT4). The 4 yields are associated with the bipolar stepper engine.

The 4 data sources are associated thusly:

- IN1 to Arduino pin 8
- IN2 to Arduino pin 9
- IN3 to Arduino pin 10
- IN4 to Arduino pin 11
* The L293D is worked with 2 VCC pins: VCC1 (pin 16) and VCC2 (pin 8). VCC1 is associated with an Arduino +5V pin. VCC2 is joined to another force source (positive terminal) with voltage proportional to engine ostensible voltage, it's distinguished in the circuit outline as (V_Motor = engine voltage).
* The 10k ohm potentiometer serves to alter the speed of the stepper engine; its yield pin is associated with Arduino simple pin 0.

The press button, which is associated with Arduino, pin 4 where it modifies the turn bearing of the stepper engine. Create a new sketch and input this code;

```
#include <Stepper.h>
// change this to the number of steps on your motor
#define STEPS 20
Stepper stepper(STEPS, 8, 9, 10, 11);
Const int button = 4; // direction control button is connected to Arduino pin 4
Const int pot  = A0; // speed control potentiometer is connected to analog pin 0
```

```
Void setup()
{
    // configure button pin as input with internal pull up enabled
    pinMode(button, INPUT_PULLUP);
}


Int direction_ = 1, speed_ = 0;


Void loop()
{
    If ( digitalRead(button) == 0 ) // if button is pressed
      If ( debounce() ) // debounce button signal
      {
          Direction_ *= -1; // reverse direction variable
          While ( debounce() ) ; // wait for button release
      }

    // read analog value from the potentiometer
    Int val = analogRead(pot);
    // map digital value from [0, 1023] to [5, 100]
    //min speed = 5 and max speed = 100 rpm
    If ( speed_ != map(val, 0, 1023, 5, 100) )
    {// if the speed was changed
        Speed_ = map(val, 0, 1023, 5, 100);
        // set the speed of the motor
```

Bool debounce()

{

  Byte count = 0;

  For(byte I = 0; I < 5; i++) {

```
  If (digitalRead(button) == 0)
    Count++;
  Delay(10);
 }
 If(count > 2)  return 1;
 Else      return 0;
}
```

CHAPTER NINE

9.1. Playing Tones

This project describes how to go about the function, tone(), to generate notes. It plays a sound or melody when you run it

Hardware Required;

- Arduino board
- Piezo buzzer or a speaker
- Hook-up wires

Circuit setup;

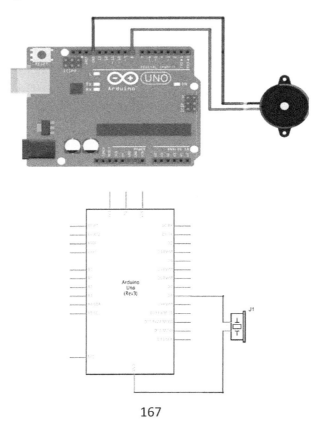

Here's the main sketch;

```
*/
Const int speakerPin = 9;   // connect speaker to pin 9
Const int pitchPin = 0;     // pot that will determine the frequency of the tone
Void setup()
{
}
Void loop()
{
  Int sensor0Reading = analogRead(pitchPin);   // read input to set frequency
  // map the analog readings to a meaningful range
  Int frequency  = map(sensor0Reading, 0, 1023, 100,5000); // 100Hz to 5kHz
  Int duration = 250;
  Tone(speakerPin, frequency, duration); // play the tone
  Delay(1000); // pause one second  82
```

Upload the sketch to your Arduino and run it.

9.2. Generating More Than One Simultaneous Tone

Here, the task is to play two tones at the same time. The Arduino Tone library only gives out a single tone on a standard board, and you desire two simultaneous tones. Bear in mind that the Mega board has extra timers and can produce up to six tones.

```
#include <Tone.h>

Int notes[] = { NOTE_A3,
                NOTE_B3,
                NOTE_C4,
                NOTE_D4,
                NOTE_E4,
                NOTE_F4,
                NOTE_G4 };
// You can declare the tones as an array
Tone notePlayer[2];
Void setup(void)
{
  Serial.begin(9600);
  notePlayer[0].begin(11);
  notePlayer[1].begin(12);
}
Void loop(void)
{
  Char c;
  If(Serial.available())
  {
```

```
C = Serial.read();
Switch(c)
{
  Case 'a'…'g':
    notePlayer[0].play(notes[c – 'a']);
    Serial.println(notes[c – 'a']);
    Break;
  Case 's':
    notePlayer[0].stop();
    break;
  case 'A'…'G':
    notePlayer[1].play(notes[c – 'A']);
    Serial.println(notes[c – 'A']);
    Break;
  Case 'S':
    notePlayer[1].stop();
    break;
  default:
    notePlayer[1].stop();
    notePlayer[0].play(NOTE_B2);
    delay(300);
```

```
        notePlayer[0].stop();

        delay(100);

        notePlayer[1].play(NOTE_B2);

        delay(300);

        notePlayer[1].stop();

        break;
    }
  }
}
```

9.3. Controlling MIDI

MIDI means Musical Instrument Digital Interface. It's a method utilized to interface devices that generate and control sound -specifically synthesizers, samplers, and computers — to make them interact with each other, via MIDI messages.

In this project, we play music on Arduino using the MIDI synthesizer. Hardware required are;

- An Arduino board
- A 220-ohm resistor
- MIDI connector
- Five-pin DIN plug

To connect to a MIDI device, a five-pin DIN plug or socket will be needed. If you use a socket, then you will likewise use a lead to connect to the device. Interface the MIDI connector with the Arduino using a 220-ohm resistor, as indicated below

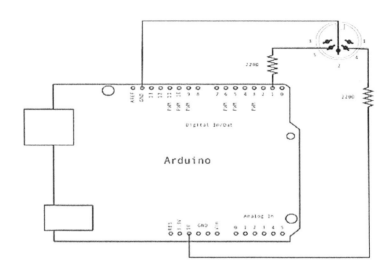

When it's time to upload the code onto Arduino, you should detach the MIDI device, as it may disrupt the upload. After uploading the sketch, connect a MIDI sound device to the Arduino output. A musical scale will play each time you press the button connected to pin 2:

//these numbers specify which note

Const byte notes[8] = {60, 62, 64, 65, 67, 69, 71, 72};

//they are part of the MIDI specification

Const int length = 8;

Const int switchPin = 2;

Const int ledPin = 13;

```
Void setup() {

  Serial.begin(31250);

  pinMode(switchPin, INPUT);

  digitalWrite(switchPin, HIGH);

  pinMode(ledPin, OUTPUT);

}

Void loop() {

 If (digitalRead(switchPin == LOW))

 {

   For (byte noteNumber = 0; noteNumber < 8; noteNumber++)

   {

    playMidiNote(1, notes[noteNumber], 127);

    digitalWrite(ledPin, HIGH);

    delay(70);

    playMidiNote(1, notes[noteNumber], 0);

    digitalWrite(ledPin, HIGH);

    delay(30);

   }

 }

Void playMidiNote(byte channel, byte note, byte velocity)

{
```

```
Byte midiMessage= 0x90 + (channel – 1);

Serial.write(midiMessage);

Serial.write(note);

Serial.write(velocity);

}
```

CHAPTER TEN

10.1 Responding to an infrared remote control

In this project, we will interface an infrared remote control with the Arduino and afterwards use it to control certain functions, such as controlling LEDs. The IR sensor is a 1838B IR receiver. Whenever you press a button on the remote, it will send an infrared signal to the IR sensor in the coded form. The IR sensor will then receive this signal and relay it to the Arduino.

Circuit setup

- To start with, connect the four LEDs to the Arduino. Attach the positive terminals of the four LEDs to the pins 7, 6, 5, and 4. Connect the negative terminals of the four LEDs to GND on the Arduino through the 220-ohm resistors. The longer wires on the LEDs are positive and the shorter wires are negative. Afterwards, connect the IR sensor to the Arduino. The connections for the IR sensor with the Arduino should be done this way;
  - The negative wire on the IR sensor should be connected to GND on the Arduino.
  - Attach the middle of the IR sensor, which is the VCC to 5V on the Arduino.
  - The signal pin on the IR sensor should be attached to pin 8 on the Arduino.

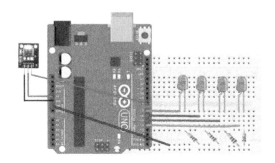

```
#include

#define first_key 48703

#define second_key 58359

#define third_key 539

#define fourth_key 25979

Int receiver_pin = 8;

Int first_led_pin = 7;

Int second_led_pin = 6;

Int third_led_pin = 5;

Int fourth_led_pin = 4;

Int led[] = {0,0,0,0};

IRrecv receiver(receiver_pin);

Decode_results output;

Void setup()
{
Serial.begin(9600);

Receiver.enableIRIn();

pinMode(first_led_pin, OUTPUT);

pinMode(second_led_pin, OUTPUT);

pinMode(third_led_pin, OUTPUT);

pinMode(fourth_led_pin, OUTPUT);
}
Void loop() {
If (receiver.decode(&output)) {
```

```
Unsigned int value = output.value;

Switch(value) {

Case first_key:

If(led[1] == 1) {

digitalWrite(first_led_pin, LOW);

led[1] = 0;

} else {

digitalWrite(second_led_pin, LOW);

led[2] = 0;

} else {

digitalWrite(second_led_pin, HIGH);

led[2] = 1;

}

Break;

Case third_key:


If(led[3] == 1) {

digitalWrite(third_led_pin, LOW);

led[3] = 0;

} else {

digitalWrite(third_led_pin, HIGH);

led[3] = 1;

}

Break;

Case fourth_key:


If(led[4] == 1) {

digitalWrite(fourth_led_pin, LOW);
```

```
led[4] = 0;
} else {
digitalWrite(fourth_led_pin, HIGH);
led[4] = 1;
}
Break;
}
Serial.println(value);
Receiver.resume();
}
}
```

10.2. Controlling a Digital Camera

What we'll be doing basically is to trigger the shutter of the camera, to do that we need to place a 2.2 k Ohms resistor between the deep parts of the poles on the 2.2 mm jack.

We will control the insertion of the 2.2 k Ohm resistor with an optocoupler. This could be carried out as a form of relay. This "trigger circuit" will be overseen by a regular output on an Arduino. Additionally, there's a LG-JG20MA sensor connected to the Arduino. This sensor triggers the Arduino.

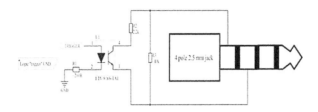

```
#define SENSOR_PIN 3
#define TRIGGER_PIN 2
Void setup() {
  pinMode(TRIGGER_PIN, OUTPUT);
  digitalWrite(TRIGGER_PIN, LOW);
  pinMode(SENSOR_PIN, INPUT_PULLUP);
  Serial.begin(9600);
}
Void camera_trigger(){
  digitalWrite(TRIGGER_PIN, HIGH);
  delay(100);

  digitalWrite(TRIGGER_PIN, LOW);
}
Uint8_t read_sensor(){
  Return digitalRead(SENSOR_PIN);
}

Void loop() {
  If(!read_sensor()){
    Delay(270); //
    Camera_trigger();
    Delay(2000); //
  }
}
```

10.3. Using millis to determine duration

A basic method of implementing timing is to make a schedule and keep an eye on the clock. Rather than world-stopping delay, you just check the clock from time to time so you know when it is time to act. Nevertheless, the processor is still free for other tasks to carry out their functions. A very simple example of this is the Blink-WithoutDelay example sketch that comes with the IDE.

```
// set pin numbers:
Const int ledPin = 13;   // the number of the LED pin


// Variables will change:
Int ledState = LOW;      // ledState used to set the LED
Long previousMillis = 0;   // will store last time LED was updated
// the follow variables is a long because the time, measured in miliseconds,
// will quickly become a bigger number than can be stored in an int.
Long interval = 1000;    // interval at which to blink (milliseconds)

Void setup() {
    // set the digital pin as output.
    pinMode(ledPin, OUTPUT);
}
Void loop()
    Unsigned long currentMillis = millis();
    If(currentMillis - previousMillis > interval) {
        // save the last time you blinked the LED
        previousMillis = currentMillis;
        // if the LED is off turn it on and vice-versa:
        if (ledState == LOW)
          ledState = HIGH;
        else
          ledState = LOW;
        // set the LED with the ledState of the variable:
        digitalWrite(ledPin, ledState);
    }
```

The wiring for this sketch is shown below;

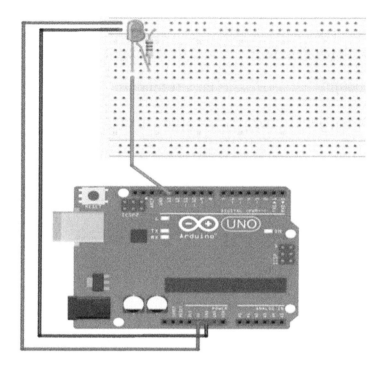

10.4. Creating Pauses in your sketch

The delay() function is used to input pauses into programs. This program pauses the sketch for the amount of time (in milliseconds) as stated in the command.

Int ledPin = 13;   // LED connected to digital pin 13

Void setup()

{

  pinMode(ledPin, OUTPUT);   // sets the digital pin as output

}

```
Void loop()
{
  digitalWrite(ledPin, HIGH);   // sets the LED on
  delay(1000);                  // waits for a second
  digitalWrite(ledPin, LOW);    // sets the LED off
  delay(1000);                  // waits for a second
}
```

10.5. Arduino as a clock

Required components;

- 1 Arduino
- 1 16X2 LCD
- 1 DS3231 RTC Module
- 1 Buzzer
- 1 10K potentiometer
- 1 220-ohm resistor
- Connecting wires
- 1 Breadboard

This project will put us through with the process of making an Arduino alarm clock using the DS3231 real time clock module. This module is affordable and operates through I2C communication, which makes it easy to incorporate with the microcontrollers. With this module, we'll get the time and cause the buzzer to beep after comparing the current time with the alarm time.

Circuit Diagram

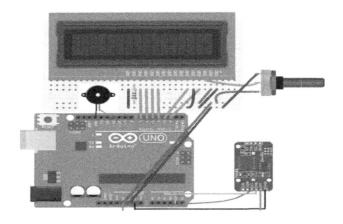

Set up the connection this way;

- Pin 1 on the LCD should be interfaced with ground on the Arduino.
- Connect Pin 2 on the LCD to 5V on the Arduino.
- Pin 3 on the LCD should be connected to the middle pin on the 10K potentiometer.
- Attach Pin 4 on the LCD to digital pin 2 on the Arduino.
- Pin 5 on the LCD to the ground of Arduino. This will put the LCD in read mode.
- Pin 6 on the LCD to the pin 3 of Arduino.
- Connect the data pins (D4-D7) to the pins 4, 5, 6, 7 on the Arduino.
- Attach Pin 15 to the 5V pin on the Arduino through the 220-ohm resistor. This is the positive pin of the backlight. Then, connect Pin 16 to ground on the Arduino. This is the negative pin of the backlight.

- Afterwards, connect the DS3231 real time clock module to the Arduino as follows:
- The GND on the DS3231 should be connected to the GND on the Arduino
- VCC on the DS3231 should be interfaced with the 5V pin on the Arduino
- SCL on the DS3231 to A5 on the Arduino
- SDA on the DS3231 to A4 on the Arduino
- To cap it all, connect the positive of the buzzer to pin 11 on the Arduino and the negative the of buzzer to GND on the Arduino.

```
#include <DS3231.h>

#include <Wire.h>

#include <LiquidCrystal.h>

LiquidCrystal lcd(2, 3, 4, 5, 6, 7);

DS3231  rtc(SDA, SCL);

Time  t;

#define buz 11

Int Hor;

Int Min;

Int Sec;

Void setup()

{
```

```
Wire.begin();

Rtc.begin();

Serial.begin(9600);

pinMode(buz, OUTPUT);

lcd.begin(16,2);

lcd.setCursor(0,0);

lcd.print("DIYHacking.com");

lcd.setCursor(0,1);

lcd.print("Arduino Alarm ");

// The following lines can be uncommented to set the date and time

//rtc.setDOW(WEDNESDAY);    // Set Day-of-Week to SUNDAY

//rtc.setTime(12, 0, 0);    // Set the time to 12:00:00 (24hr format)

//rtc.setDate(1, 1, 2014);  // Set the date to January 1st, 2014

  Delay(2000);

}
Void loop()

{

  T = rtc.getTime();

  Hor = t.hour;

  Min = t.min;
```

```
Sec = t.sec;

Lcd.setCursor(0,0);

Lcd.print("Time: ");

Lcd.print(rtc.getTimeStr());

Lcd.setCursor(0,1)

Lcd.print("Date: ")

Lcd.print(rtc.getDateStr());

If( Hor == 11 &&  (Min == 32 || Min == 33)) //Comparing the current time with the Alarm time

Buzzer();

Buzzer();

Lcd.clear();

Lcd.print("Alarm ON");

Lcd.setCursor(0,1);

lcd.print("Alarming");

Buzzer();

Buzzer();

}

 delay(1000);

}

void Buzzer()
```

{

digitalWrite(buz,HIGH);

delay(500);

digitalWrite(buz, LOW);

delay(500);

}

10.5. Using a real-time clock

The Time Clock Module (or DS3231) is a module that gives precise estimates of the time, either through the Arduino card cell or independently. The Arduino card measures the time used up starting from when the module was turned on (in ms).

Required components;

- An Arduino board
- Jumper wires
- Time clock module zs-042

Circuit connection

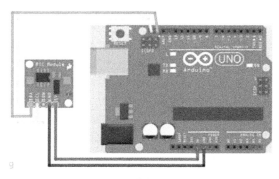

Code;

```c
#include <Wire.h>
#include <ds3231.h>
struct ts t;
void setup() {
  Serial.begin(9600);
  Wire.begin();
  DS3231_init(DS3231_INTCN);
  /*Insert these values to synchronize clock module/
  t.hour=12;
  t.min=30;
  t.sec=0;
  t.mday=25;
  t.mon=12;
  t.year=2019;

  DS3231_set(t);
}

void loop() {
```

```
DS3231_get(&t);
Serial.print("Date : ");
Serial.print(t.mday);
Serial.print("/");
Serial.print(t.mon);
Serial.print("/");
Serial.print(t.year);
Serial.print("\t Hour : ");
Serial.print(t.hour);
Serial.print(":");
Serial.print(t.min);
Serial.print(".");
Serial.println(t.sec);
 delay(1000);
}
```

CHAPTER ELEVEN

11.1. Connecting to an Ethernet network

Connecting an Arduino to an Ethernet network grants it online accessibility. We'll use an Arduino Ethernet Shield here. The Ethernet shield will then be stacked above the Arduino via a Shield interface.

These are the procedures to follow:

- Fix the Ethernet Shield tightly on the Arduino hardware as shown in the image

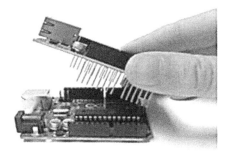
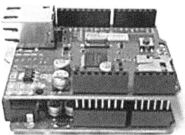

- Insert the Ethernet Shield to a network router, or to your computer, via an RJ45 cable.

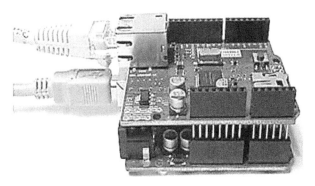

11.2. Using Arduino as a webserver

When you interface Arduino with an Ethernet shield, it can be used as a simple web server, and by entering that server with a browser operating on any computer connected to the same network as the Arduino, you can perform numerous operations such as reading the value of a sensor.

Required components;

- Arduino board
- Ethernet shield
- Wired LAN Connection with a speed of 10/100Mb
- An Ethernet cable
- Wi-Fi Router
- A Breadboard
- 3 Jumper Wires
- 10k Resistor
- Two 9V Adaptor
- 1 pushbutton

Circuit setup

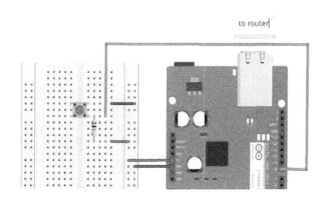

Connect the listed hardware parts as indicated above. Arduino's pin 8 is connected to the pushbutton and is programmed as INPUT. When you push the button, the Arduino will read a LOW value on this pin. The Arduino will then convert the status of the OUTPUT to ON. The output will be set to OFF when it is released. The status of the switch will be transmitted to the Web server.

Afterwards, specify the IP address and MAC address by inputting these lines;

Byte ip[] = { 192, 168, 0, 112 };

Byte mac[] = { 0x90, 0xA2, 0xDA, 0x0D, 0x85, 0xD9 };

Code:

Client.println("<!DOCTYPE html>"); //web page is made using HTML

Client.println("<html>");

Client.println("<head>");

```
Client.println("<title>Ethernet Tutorial</title>");

Client.println("<meta http-equiv=\"refresh\" content=\"1\">");

Client.println("</head>");

Client.println("<body>");

Client.println("<h1>A Webserver Tutorial </h1>");

Client.println("<h2>Observing State Of Switch</h2>");

Client.print("<h2>Switch is:  </2>");

If (digitalRead(8))

{

Client.println("<h3>ON</h3>");

}

Else

{

Client.println("<h3>OFF</h3>");

}

Client.println("</body>");

Client.println("</html>");
```

This sketch will lead to a web page on a Web browser when the IP address assigned to the Arduino is accessed.

11.3. Sending Twitter messages

The objective is to demonstrate how we can use Arduino to post automatic and cyclical tweets to share the status of the Analog inputs on Twitter.

Required components;

- An Arduino board
- An Arduino Ethernet shield

Code;

```
#include <SPI.h> // needed in Arduino 0019 or later

#include <Ethernet.h>

#include <Twitter.h>

// Ethernet Shield Settings

Byte mac[] = { 0xDE, 0xAD, 0xBE, 0xEF, 0xFE, 0xED };

// Message to post

Char msg[200];

Int a0,a1,a2,a3,a4,a5;

Int n=0;

Void setup()
{
  Delay(1000);
  Ethernet.begin(mac, ip);
```

```
  Serial.begin(9600);
    Serial.println("connecting ...");
}
Void loop()
{
While(1){ //repeat for ever
        N++; // Index to say which order of tweet is
        //get some data from Aanalog inputs.
        A0=analogRead(0);
        A1=analogRead(1);
        A2=analogRead(2);
        A3=analogRead(3);
        A4=analogRead(4);
        A5=analogRead(5);
           Serial.println(msg);  //for testing
        If (twitter.post(msg)) { //To post the Tweet
                Int status = twitter.wait(&Serial);
                If (status == 200) {
                  Serial.println("OK.");
                } else {
```

```
                Serial.print("failed : code ");

                Serial.println(status);

            }

        } else {

            Serial.println("connection failed.");

        }

    Delay(70000); //

    }

}
```

11.5. Publishing Data to an MQTT broker

MQTT is a typical transfer protocol directed at small IoT enabled devices. While the Arduino has no networking capability independently, it can be interfaced with an Ethernet shield, enabling it to connect to the internet. Using the Ethernet and MQTT library, we can easily make our Arduino talk to MQTT servers to submit and retrieve data!

The first approach to using MQTT is to describe certain variables, including the IP address, MAC, server, and certain subjects. The first line in our program is a blueprint of the function that will regulate incoming messages. The preceding lines create our MAC address, which has to be unique, the IP address of our Ethernet.

The next line states the MQTT broker that we will connect to. A broker in the MQTT realm is a server, however, in contrast to servers, brokers can send messages to clients at any time and they are not built to store data, they only transmit it.

```
// Function prototypes

Void subscribeReceive(char* topic, byte* payload, unsigned int length);

// Set your MAC address and IP address here

Byte mac[] = { 0xDE, 0xAD, 0xBE, 0xEF, 0xFE, 0xED };

IPAddress ip(192, 168, 1, 160);

// Make sure to leave out the http and slashes!

Const char* server = "test.mosquitto.org";

// Ethernet and MQTT related objects

EthernetClient ethClient;

PubSubClient mqttClient(ethClient);

Void setup()
{
  // Useful for debugging purposes
  Serial.begin(9600);
  // Start the ethernet connection
  Ethernet.begin(mac, ip);
```

```
// Ethernet takes some time to boot!
Delay(3000);
// Set the MQTT server to the server stated above ^
mqttClient.setServer(server, 1883);
// Attempt to connect to the server with the ID "myClientID"
If (mqttClient.connect("myClientID"))
{
  Serial.println("Connection has been established, well done");
  // Establish the subscribe event
  mqttClient.setCallback(subscribeReceive);
}
Else
{
  Serial.println("Looks like the server connection failed…");
}
}
```

11.6. Using built-in Libraries

Libraries in Arduino are files composed in C or C++, which provide you with additional functionalities in your sketches. For example, reading an encoder, controlling an LED.

To utilize a built-in library in a sketch, just move to the Sketch menu, click "Import Library", and select from the libraries that are displayed. This will bring in an #include statement at the crest of the sketch for each header (.h) file in the library's folder.

11.7. Installing a third-party library

The Library Manager can be used to install a third-party library into your Arduino IDE. Launch the IDE and move to the "Sketch" menu and then Include Library > Manage Libraries.

The Library Manager will provide you with a list of libraries that are already installed or about to be installed will be displayed. For instance, let's install the Bridge library. Scroll the list to find it, click on it when you see it. Then, choose the version of the library you intend to install. If it doesn't display the version selection menu, don't panic, there's no issue.

To wrap it all up, press install and hold on for the IDE to install the new library. The download process may consume time contingent upon your connection speed. Upon completion, an Installed tag should appear close to the Bridge library. If you want to access the library, press Sketch > Include Library menu.

CHAPTER TWELVE

12.1. Arduino Build Process

The processes involved when you want to build a sketch in the Arduino. Various things need to occur for your Arduino code to get onto the Arduino board. To start with, the Arduino development program plays out some minor pre-processing to transform your sketch into a C++ program. Next, the conditions of the sketch are found. It is at that point sent to a compiler (Avr-GCC), which transforms the human-readable code into machine-readable code. At that point, your code is linked against the standard Arduino libraries that give essential functions like digitalWrite() or Serial.print(). The outcome is a solitary Intel hex document, which contains the particular bytes that should be written in the program memory of the chip on the Arduino board. This file is then transferred to the board: communicated over the USB or sequential association through the bootloader present the chip or with outside programming hardware.

12. 2. Measuring Used RAM on Arduino

In the Arduino IDE directory where you'll find the avr-gcc compiler, there's also a tool called 'avr-size'. Move to hardware/tools/avr/bin/, you'll find this tool there. The IDE makes an impermanent catalog in your temp index when assembling, and duplicates the whole C(++) documents to it.

```
C:\Users\Hans>D:\arduino\hardware\tools\avr\bin\avr-size.exe C:\Users\Hans\AppDa
ta\Local\Temp\build284110048787601542.tmp\Series1_Ix.cpp.elf
   text    data     bss     dec     hex filename
   3924      32     320    4276    10b4 C:\Users\Hans\AppData\Local\Temp\build28
4110048787601542.tmp\Series1_Ix.cpp.elf

C:\Users\Hans>
```

The well-established outcomes are like this. The outcome implies;

Text – streak information utilized for code

Information – Memory with introduced information

Bss – Memory that is introduced with zero's (the compiler will incorporate some code so it will instate information and bss)

This is a typical apparatus for fixed RAM distributions, later on, it's imperative to endure at the top of the priority list that it doesn't check the RAM utilized via programmed factors neighborhood to a capacity or square which are designated on the stack toward the beginning of execution of that work, that must be found out by actualizing a specific channel through the program.

12.3. Store and Retrieve Values

Variables represent names you can create for storing, retrieving, and using values in the Arduino's memory. Let's take a look at these variables for instance;

Int a = 42;

Char c = 'm';

Float root2 = sqrt(2.0);

The statement int a = 42 creates a variable termed a. The int informs the Arduino software about the type of variable it's handling. The int type can store integer values ranging from -32,768 to 32,767. Then, char c = 'm' gives a variable named c of the type char (which is for storing characters) and then assigns it the value 'm'. Float root2 = sqrt(2.0) provides a variable named root2. The variable type is float, and it holds decimal values. In this sketch, root2 is initialized to the floating-point representation of the square root of two: sqrt(2.0). Now that the values have been stored in the memory, let's figure out how to retrieve them. A method of doing this is to simply pass each variable to a function's parameter which is by using the Serial.println(Val) function which displays the value of the variable inside the bracket.

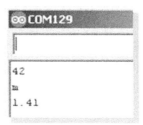

```
Serial.println(a);
Serial.println(c);
Serial.println(root2);
```

```
42
m
1.41
```

Void setup()

{

  Serial.begin(9600);

  Int a = 42;

  Char c = 'm';

  Float root2 = sqrt(2.0);

  Serial.println(a);

  Serial.println©;

  Serial.println(root2);

}

Void loop()

{

  // Empty, no repeating code.

}

12.4. Storing Data permanently on EEPROM memory

A number of Arduino boards give users the credibility to permanently store data in an EEPROM without plugging in the board. EEPROM means Electronically Erasable Programmable Read-Only Memory. Even though the data on the chip can be overwritten, there is a limitation to the number of times that these can be done. Else, it might start malfunctioning. Nevertheless, you can read for as long and as many times and you intend to.

You can store data permanently on the EEPROM by using The Write () method. The sketch connotes how you can store a byte using the write () method

```
#include <EEPROM.h>

Void setup(){

    Int word_address = 0;

    EEPROM.write(word_address, 0x7F);

}

Void loop()

{}
```

Incorporate the write() method with a word address and the value you intend to store. The address should be a value between zero and EEPROM.length()-1 and it instructs the MCU where to store the value.

12.5. Periodic Interrupt

This is a method of performing periodic actions on Arduino by programming specific actions to be done at specific times. The priority is to include an interrupt handler in your program, it should be inputted above the Setup () line.

-const unsigned long TIMER0_COUNT = 500;   // 500 msec timer interval

// TIMER0 interrupt handler

Volatile bool time = false;

ISR(TIMER0_COMPA_vect) {

  Static unsigned long count = 0;

  If (++count > TIMER0_COUNT) {

    Count = 0;

    Time = true;

  }

}

Next, we set the time interval. This routine appends itself onto the TIMER0 interrupt, which is set to initiate at approximately every 1 msec. Your "interval" denotes the number of TIMER0 interrupts to process. Each interval is approximately 1 millisecond, therefore, you're basically setting how many TIMER0 interrupts to count before activating your interval. IOW, inputs the variable TIMER0_COUNT to the number of milliseconds you intend to wait. For instance, if you want to wait for 5 seconds, use 5000

Afterwards, you enter the "TIMER0 initialization" code to your setup() method.

```
// *** TIMER0 initialization ***
    Cli();                      // disable all interrupts
    TIMSK0 = 0;                 // disable timer0 for lower jitter
    OCR0A = 0xBB;               // arbitrary interrupt count
    TIMSK0 |= _BV( OCIE0A );    // append onto interrupt
    Sei();                      // activates interrupt
```

Now, you just need to include the "time check" code to your loop() method.

```
If ( time ) {
  Time = false;
  // do something here
}
```

The "time = false;" line is essential. If this line is absent, the "do something here" line(s) would be executed every time the program carries out the loop() function. Definitely, you'll replace activities in the "do something here" line by printing some text or flashing the LED.

12.6. Changing a Timer's PWM frequency

One of the basic applications of the Arduino Uno is in high-frequency circuits. For a controller to be used in a high-frequency circuit like in a buck converter, the controller must be capable of generating high-frequency PWM waves. With the Arduino Uno as your controller, you must know how to change the frequency on PWM pins of Arduino Uno. PWM stands for pulse width modulation.

Here is the default frequency of each PWM pin of Arduino UNO:

PWM frequency for D3 & D11:

490.20 Hz (The DEFAULT)

PWM frequency for D5 & D6:

976.56 Hz (The DEFAULT)

PWM frequency for D9 & D10:

490.20 Hz (The DEFAULT)

Now, these frequencies are ideal for low-frequency applications like fading an LED. However, these default frequencies are not compatible with High-frequency circuits; hence, we need to change them.

Let's take a look at the code for the D3 & D1

```
TCCR2B = TCCR2B & B11111000 | B00000001; // for PWM frequency of 31372.55 Hz
TCCR2B = TCCR2B & B11111000 | B00000010; // for PWM frequency of 3921.16 Hz
TCCR2B = TCCR2B & B11111000 | B00000011; // for PWM frequency of 980.39 Hz
TCCR2B = TCCR2B & B11111000 | B00000100; // for PWM frequency of 490.20 Hz (The DEFAULT)
TCCR2B = TCCR2B & B11111000 | B00000101; // for PWM frequency of 245.10 Hz
TCCR2B = TCCR2B & B11111000 | B00000110; // for PWM frequency of 122.55 Hz
TCCR2B = TCCR2B & B11111000 | B00000111; // for PWM frequency of 30.64 Hz
```

The Arduino circuit is imitated in duplicity in order to show how the changes occur when you apply the code.

- Select two Arduino Uno and place them on Front-Panel
- Connect the digital Pin 3 (PWM pin) of each Arduino to the oscilloscope

You'll then write two separate programs for these Arduinos; Program A – Default frequency on Pin 3

```
Program A – Default frequency on Pin 3
Void setup() {
pinMode(3,OUTPUT);
// put your setup code here, to run once:
}
Void loop() {
analogWrite(3,155);
// put your main code here, to run repeatedly:
}

Program B – Changed frequency on Pin 3
Void setup() {
```

```
TCCR2B = TCCR2B & B11111000 | B00000001; // for PWM frequency of 31372.55 Hz
pinMode(3,OUTPUT);
// put your setup code here, to run once:
}
Void loop() {
analogWrite(3,155);
// put your main code here, to run repeatedly:
}
```

The hex file of these programs is made available to Arduino. Then, you run simulation.

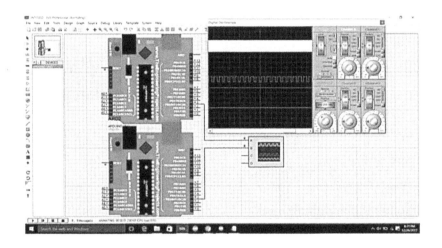

From the oscilloscope, you will see that frequency is increased to a very high value when this line of code is inputted;

TCCR2B = TCCR2B & B11111000 | B00000001; // for PWM frequency of 31372.55 Hz

12.7. Counting Pulses

In this task, we will go through how an Arduino Uno can be used to count pulses, or signals, from the sensor utilizing an interrupt and execute functions. This has been programmed as RISING, therefore, it counts the pulses from digital signal zero to digital signal one:

```
Int pin = 2;

Volatile unsigned int pulse;

Constintpulses_per_litre = 450;

Void setup()
{
Serial.begin(9600);
pinMode(pin, INPUT);
attachInterrupt(0, count_pulse, RISING);
}
Void loop()
{
Pulse=0;
Interrupts();
Delay(1000);
noInterrupts();
```

Serial.print("Pulses per second: ");

Serial.println(pulse);

}

Voidcount_pulse()

{

Pulse++;

}

Launch the Arduino Serial Monitor, and blow air into the water flow sensor through your mouth. The amount of pulses per second will be displayed on the Arduino Serial Monitor for each loop, as indicated in the image;

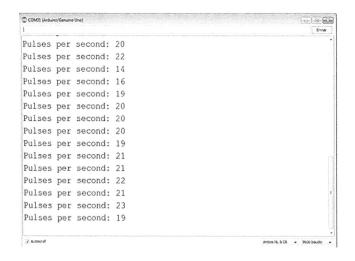

12.8. Reducing Battery Drain

The Arduino UNO is built as a development board, rather than a particular low power gadget, it has numerous design models. This implies that it consumes more power than the necessary minimum. If you want to lessen the general battery consumption of the Arduino UNO board;

- Supplant the linear regulator with a DC-DC regulator.
- Modify the USB-to-Serial circuit so it's powered from the USB port,
- Detach (or unsolder) the consistently-on Light Emitting Diodes on the board,
- Utilize the processor sleep mode.

Through on-board circuit changes, specific substitution of components and the use of microcontroller rest mode can lessen the persistent idle power usage of an Arduino UNO.

12.9. Uploading Sketches using a programmer

First thing to do is to connect the programmer to your Arduino. After that, open up the IDE. Then, open an example sketch like Blink (File > Examples > 1.Basics > Blink).

Prior to uploading, you should inform Arduino about the programmer you're using. Go up to Tools > Programmer and select USBtinyISP.

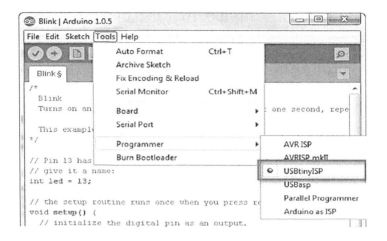

Furthermore, ensure you set the Board option in the correct manner. The serial port selection isn't compulsory for uploading the sketch, however, it is still necessary if you're doing anything that involves the serial monitor. To upload the sketch via the programmer you chose, go to File > Upload Using Programmer. If you'll be doing this severally, it can easily be done by pressing CTRL+SHIFT+U (COMMAND+SHIFT+U on Mac).

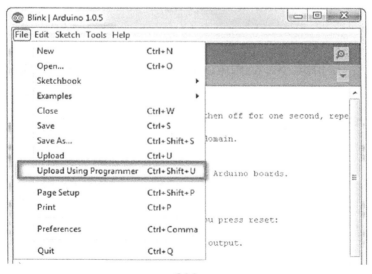

12.10. Replacing Arduino Bootloader

If you've uploaded a sketch via the programmer, automatically you've also erased the Arduino bootloader. Hence, we have to replace the serial bootloader on your Arduino. One of the embedded features of the Arduino IDE is to allow users to (re-)upload a bootloader to the AVR. Here's how:

Ensure you've set the Board option correctly as stated earlier. Then, move to Tools > Burn Bootloader at the base of the menu. This process may delay a bit because asides from writing the bootloader into the flash of your AVR, the fuse bits (setting the clock speed, bootloader space, etc), and lock bits (barring the bootloader from overwriting itself) will also be (re)set.

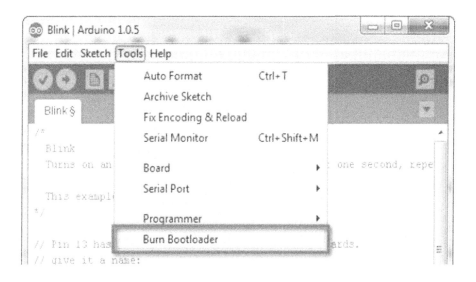

The bootloader upload process is said to be complete when the "Burning bootloader to I/O board (this may take a minute)..." message becomes "Done burning bootloader."

Overall, the Arduino microcontroller is a versatile gadget, very versatile, and it fits conveniently into our daily digital uses. Have a wonderful time exploring this superb gadget!

About the Author

Obakoma G. Martins is a tech enthusiast with several years of experience in the ICT industry. He is a geek and passionately follows the latest technical and technological trends. His strength lies in figuring out the solution to complex tech problems. Obakoma holds a Bachelor's Degree in Computer Science.

Made in the USA
Monee, IL
15 December 2021